Location

Maps, plans and grids

A1 (a) Bob is holding the map in his right hand.
(b) Julie is pointing with her left hand.
(c) David has the lead in his right hand.
(d) The dog is scratching its left ear.
(e) Sandra is holding Peter's right hand.
(f) Alan, who is doing a cartwheel, has his left hand on the ground.

A2 (a) If you stood by the church and looked at the farm, the big tree would be to your left.
(b) If you stood by the big tree and looked at the windmill, the church would be to your right.
(c) If you stood by the farm and looked towards the church, the windmill would be to your left.

A3 If you start from the railway bridge and go along Wood Street then:
(a) Ash Way is 2nd left
(b) Beech Avenue is 1st right
(c) Poplar Walk is 3rd right
(d) Chestnut Hill is 5th left
(e) Aspen Lane is 4th right.

A4 If you start at the clock tower and go along Wood Street towards the railway bridge, then:
(a) the first turning on the left is Aspen Lane
(b) Birch Grove is the third right
(c) Beech Avenue is 4th left.

A5 If you go along Elm Gardens towards Wood Street, you need to turn right to get to the clock tower.

A6 If you are walking towards Wood Street from Birch Grove, you need to turn right to get to the railway bridge.

A7

If you start at the pond and walk along Windy Lane then:
(a) the first turning on your left is Amber Avenue. (not Bolter Hill – if you're not sure why, ask your teacher)
(b) Downend Road is second on your right. (not Coronation Road)
(c) If you turn into Coronation Road then the post office is on the left side.
(d) If you pass the post office you turn right into Ashton Road.
(e) If you walk along Ashton Way the school is on your right.
(f) If you turn round and walk back the school is now on your left.

A8 To direct someone to Mr Scott's house from the post office you should say something like this.
'Turn left out of the post office, then take the second right, then turn first left, then turn second left. Mr Scott's house is at the end of the street.'

If your directions are different, test them on a friend – you might be right!

A9

The name of this road is Bolter Hill.

A10

The roads you can see are Ashton Way and Mill Lane.

B1

He could say: '5 from the left on the 6th floor.'
There are other ways he could say where his flat is.

B2

Gary could say: '4 rows of houses from the left, 5th house in the row.'

C1 (a) The clock tower is in B3.
▲ (b) The cinema is in C1.
(c) C4 is where the hospital is.
(d) E3 is where the zoo entrance is.

C2 The school is in square E1.

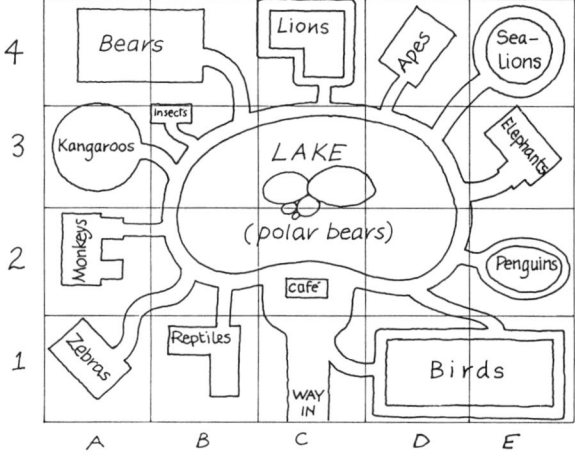

C3 Kangaroos are in the round cage.

C4 (a) Penguins are in the house marked A.
(b) Apes are in B.
(c) Insects are in the house marked C.

C5 (a) The monkey house is in square A2.
(b) If you went out of the monkey house and then turned left you would come to the kangaroos.
(c) If you walk round the lake clockwise, the zebras come after the reptile house.

C6 (a) B1 is where the reptiles are.
(b) The apes are in D4.
(c) E4 is the square which the sealions are in.
(d) In C2 you will find the café.
(e) The bird house is in D1 and E1.
(f) Zebras are in the animal house in square A1.
(g) The insect house is in square B3.

	Sun	Mon	Tue	Wed	Thur	Fri	Sat
Zoo opens	11:00am	9:30am	9:30am	9:30am	9:30am	9:30am	9:00am
Sealions fed	2:00pm	2:00pm	2:00pm	2:00pm	2:00pm	——	2:00pm
Reptiles fed	——	——	——	——	——	2:30pm	——
Penguins fed	3:00pm	3:00pm	3:00pm	3:00pm	3:00pm	3:00pm	3:00pm
Birds of prey fed	3:15pm	3:15pm	3:15pm	3:15pm	——	3:15pm	3:15pm
Lions fed	3:30pm	3:30pm	3:30pm	3:30pm	3:30pm	3:30pm	3:30pm
Zoo closes	5:00pm	6:00pm	6:00pm	6:00pm	6:00pm	6:00pm	5:30pm

D1 (a) On Wednesday the zoo opens at 9:30 a.m.
(b) Reptiles are fed on Fridays.
(c) Sealions, penguins and lions are all fed on Thursdays.
(d) If you arrived at 2:50 p.m. on a Saturday, you would have to wait 40 minutes to see the lions fed.

D2 (a) On a Sunday the zoo opens for 6 hours.
(b) On a Monday the zoo opens for $8\frac{1}{2}$ hours.
(c) On a Saturday the zoo opens for $8\frac{1}{2}$ hours.

	Male or female	Date of birth	Weight at birth	Weight now
Polar bear	female	30.11.75	400 grams	320 kilograms
Black bear	female	6.5.68	325 grams	120 kilograms
Chimpanzee	male	12.3.70	2 kilograms	72 kilograms
Kangaroo	female	9.1.72	12 grams	45 kilograms
Lion	male	21.6.69	1·5 kilograms	178 kilograms
Tortoise	male	27.7.32	30 grams	4 kilograms

D3 (a) The black bear weighed 325 grams at birth.
(b) The chimpanzee weighs 72 kilograms now.
(c) The polar bear was born in November.
(d) The lion is older than the kangaroo.
(e) The kangaroo is a 'she'.
(f) The polar bear is the heaviest of the six animals.
(g) The tortoise is the oldest of the animals.
(h) The chimpanzee weighed 2 kilograms at birth.
(i) He is now 70 kilograms heavier.

E1 and **E2** Show your drawings to your teacher.

F1 HAPPY NEXT BIRTHDAY TO YOU.
[(1, 1) is H, (0, 0) is A, (3, 2) is P and so on …]

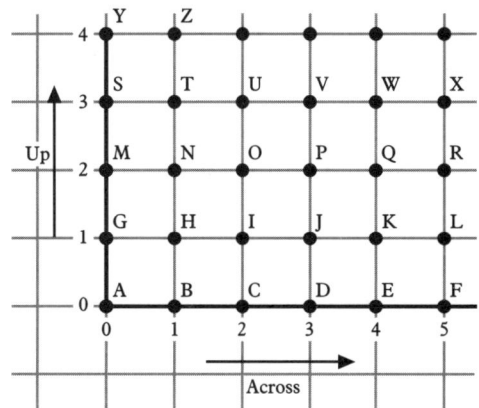

Digging into history

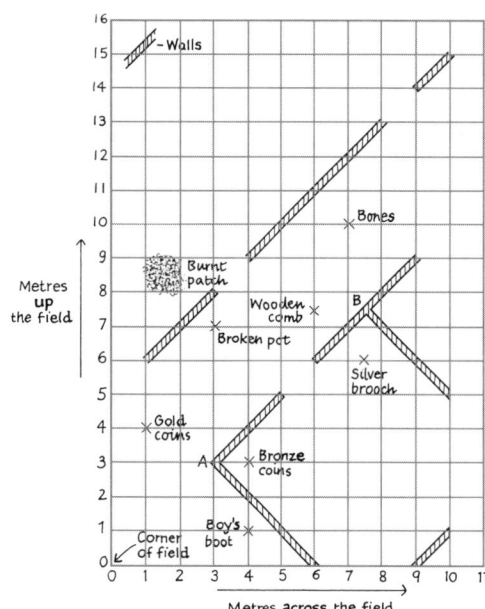

Metres **up** the field / Metres **across** the field

A1 A boy's boot was found at (4, 1). [Not gold coins – make sure you understand why.]

A2 A silver brooch was found at $(7\frac{1}{2}, 6)$.

A3 A wooden comb was found at $(6, 7\frac{1}{2})$.

A4 (a) A broken pot was found at (3, 7).
(b) Bones were found at (7, 10).
(c) Bronze coins were found at (4, 3).

A5 A wall goes from (4, 9) to (8, 13).

A6 (0, 0) is the position of one corner of the field.

A7 A is (3, 3) and B is $(7\frac{1}{2}, 7\frac{1}{2})$.

A8, A9 and **A10** Show your worksheets to your teacher.

B1 (a) 10 of Hadrian, 8 of Antoninus, 9 of Severus, 5 of Diocletian, 4 of Constantine.

B2

	Gold	Silver	Bronze	Totals
Hadrian	5	3	2	10
Antoninus	2	6	0	8
Severus	2	0	7	9
Diocletian	1	0	4	5
Constantine	0	3	1	4
Total	10	12	14	36

(b) Make sure you understand where the different totals come from – and how you use them to check your table.

B3

	Gold	Silver	Bronze	Jet
Rings	1	3	3	0
Brooches	0	2	7	0
Buckles	1	1	6	0
Pendants	1	0	4	2
Bracelets	2	1	3	1

(a) 2 silver brooches were found.
(b) 2 jet pendants were found.
(c) 6 bronze buckles were found.
(d) There were 7 rings found altogether.
(e) 23 bronze objects were found altogether.
(f) Altogether 38 pieces of jewellery were found.

B4

(a) 2·6 kg   (b) 3·8 kg

B5 (a) Room 5 had the most pottery in it (11·0 kg).
(b) This room was probably a kitchen or store room. Why?

4

C1

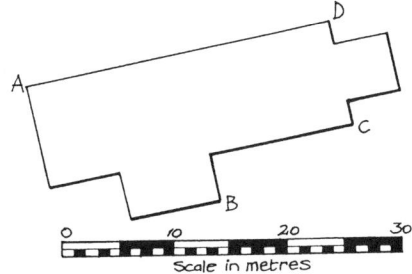

The length of AB is 20 metres, AC is 29 metres and AD is 27 metres.

C2

P •

• Q

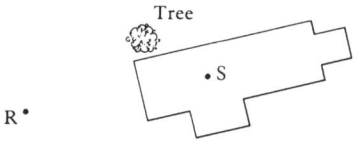

R •

• T

• U

Sketch *a* from P, sketch *b* from R, sketch *c* from T, sketch *d* from U, sketch *e* from Q, sketch *f* from S.

D1 and **D2** show your worksheet to your teacher.

Bearings and journeys

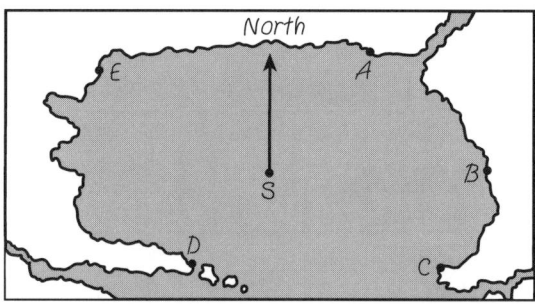

A1

Point	Bearing from S	Distance from A
A	39 to 41 degrees	3·9 to 4·1 km
B	88 to 90 degrees	5·5 to 5·7 km
C	118 to 120 degrees	4·9 to 5·1 km
D	220 to 222 degrees	2·9 to 3·1 km
E	300 to 302 degrees	4·9 to 5·1 km

A2 Show the map you have drawn to your teacher.
The biggest gap is between submarines D and E.

A3 Show worksheet 3-13 to your teacher.

A4

Coast guard station	Bearing of lighthouse	Distance of lighthouse
Shipman Head	224° to 226°	6·4 to 6·7 km
Appletree Point	12° to 14°	6·1 to 6·3 km
Brandy Bay	84° to 87°	17·4 to 17·6 km
Castle Down	110° to 112°	16·4 to 16·6 km
Toll Point	118° to 120°	8·3 to 8·5 km

A5 The distance of Toll Point from Castle Down is 8·2 to 8·4 km.
The bearing is 102° to 104°.

A6 Distance 16·7 to 16·9 km, bearing 285° to 287°

B1, B2, B3 and **B4** Show worksheet 3-14 to your teacher.

B3 The motorboat is heading for Catfish Island.

B4 The helicopter is heading for Herring Island.

B5 and **B7** Show worksheet 3-15 to your teacher.

B6

	Bearing	Distance
1st stage	83° to 85°	2·9 to 3·1 km
2nd stage	150° to 152°	3·7 to 3·9 km
3rd stage	114° to 116°	5·2 to 5·4 km
4th stage	209° to 211°	1·5 to 1·7 km
5th stage	284° to 286°	6·5 to 6·7 km
6th stage	263° to 265°	5·0 to 5·2 km
7th stage	218° to 220°	4·8 to 5·0 km
8th stage	85° to 87°	4·6 to 4·8 km

C1, **C2** and **C3** Show worksheet 3-16 to your teacher.

C3 The tanker is heading for Gull Rock.

C4

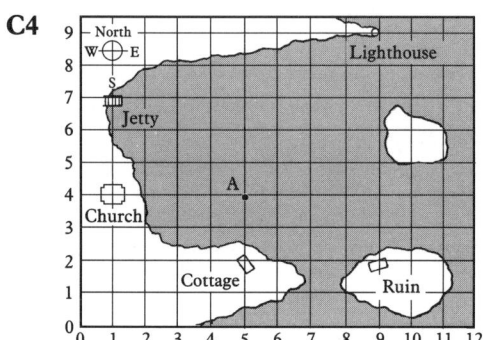

Brenda's boat is at (9, 7).

C5 (a) Colin (9, 4) (b) Diane (5, 7)
   (c) Eric (5, 6) (d) Fred (5, 8)

D1

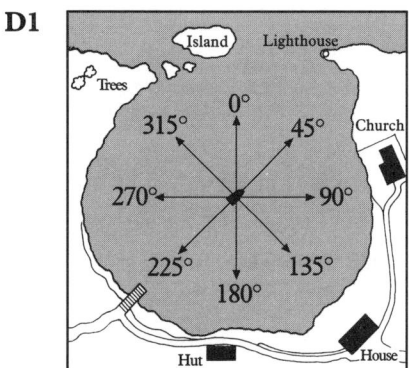

(a) 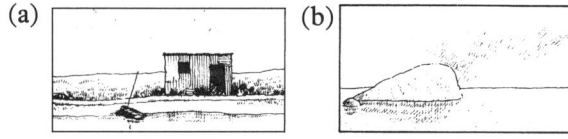 (b)

 (c)    (d)

 (e)    (f)

Picture	Bearing
(a)	180°
(b)	0°
(c)	225°
(d)	135°
(e)	315°
(f)	90°

D2

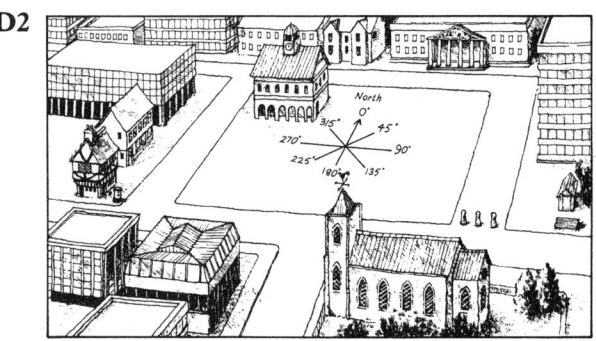

(a)  (b)  (c)

(d)  (e)  (f)

Picture	Bearing
(a)	180°
(b)	315°
(c)	45°
(d)	270°
(e)	135°
(f)	225°

Dots lines and networks

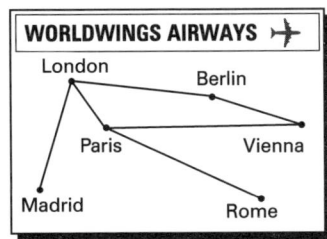

WORLDWINGS AIRWAYS ✈

A1 Here are two routes which Worldwings fly from Madrid to Rome:

        Madrid – London – Paris – **Rome**

and **Madrid** – London – Berlin – Vienna – Paris – **Rome**.

A2

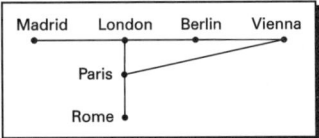

Both networks show the same connections, but the positions of some cities has been changed. This does not matter if all you are interested in is where you might have to change planes.

A3 ▲

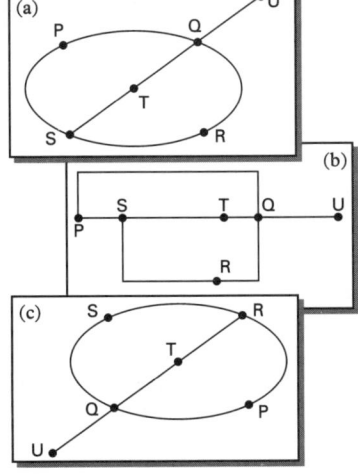

The odd network out is (c). How is it different to the others?

A4

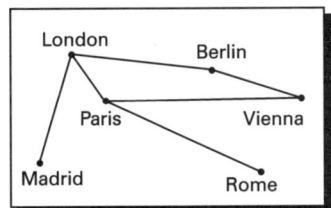

Here is the link table for the network.

	London	Paris	Rome	Berlin	Vienna	Madrid
London	0	1	0	1	0	1
Paris	1	0	1	0	1	0
Rome	0	1	0	0	0	0
Berlin	1	0	0	0	1	0
Vienna	0	1	0	1	0	0
Madrid	1	0	0	0	0	0

A5 If two networks have different link tables they are different networks. Did you find that A3(c) had a different link table to A3(a) and (b)?

A3(a)

	P	Q	R	S	T	U
P	0	1	0	1	0	0
Q	1	0	1	0	1	1
R	0	1	0	1	0	0
S	1	0	1	0	1	0
T	0	1	0	1	0	0
U	0	1	0	0	0	0

A3(b)

	P	Q	R	S	T	U
P	0	1	0	1	0	0
Q	1	0	1	0	1	1
R	0	1	0	1	0	0
S	1	0	1	0	1	0
T	0	1	0	1	0	0
U	0	1	0	0	0	0

A3(c)

	P	Q	R	S	T	U
P	0	1	1	0	0	0
Q	1	0	0	1	1	1
R	1	0	0	1	1	0
S	0	1	1	0	0	0
T	0	1	1	0	0	0
U	0	1	0	0	0	0

A6 All link tables like the ones here have reflection symmetry. This is because if there is a route between (say) P and Q there must be one between Q and P. (Think carefully about this – you may need to ask your teacher to explain.)

A7

	A	B	C	D
A	0	0	1	1
B	0	0	0	1
C	1	0	0	0
D	1	1	0	0

Here is one network which fits the link table.
Yours may be slightly different – check your own against the link table.

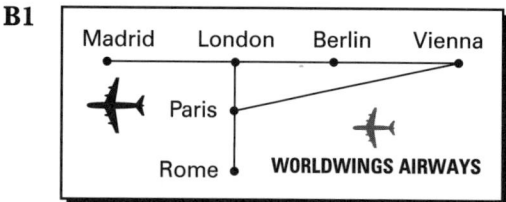

A8

	Paddington	Euston	King's Cross	Waterloo	Victoria
Paddington	0	0	1	1	1
Euston	0	0	1	1	1
King's Cross	1	1	0	1	1
Waterloo	1	1	1	0	0
Victoria	1	1	1	0	0

Anton would prefer not to change trains. So he wants a direct route.
This means that King's Cross would be the best station for him (there is a direct link between Paddington and King's Cross). Check the link table above.

A9 Some people who would be interested in direct connections are: people with very young children, people with disabilities, old people – you may have thought of some more of your own.

B1

(a) Paris has three links, so it is a degree 3 vertex.
(b) Rome only has one link. It is a degree 1 vertex.
(c) Vienna is a degree 2 vertex.

B2 (a) Here is a network with 3 vertices each of degree 2.

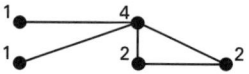

(b) It is **impossible** to make **just one** of these vertices into a degree 3 vertex. Why?

B3 (a) Sum of the degrees of all the vertices is $1 + 1 + 4 + 2 + 2 = 10$.

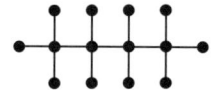

(b) Sum of the degrees of all the vertices is $(10 \times 1) + (4 \times 4) = 10 + 16 = 26$.

(c) Sum of the degrees of all the vertices is $(2 \times 2) + (2 \times 3) = 4 + 6 = 10$.

(d) Sum of the degrees of all the vertices is $(7 \times 2) + (2 \times 3) = 14 + 6 = 20$.

(e) Sum of the degrees of all the vertices is $2 + (2 \times 3) + 4 = 2 + 6 + 4 = 12$.

B4

Network	Sum of degrees of the vertices	Number of links
(a)	10	5
(b)	26	13
(c)	10	5
(d)	20	10
(e)	12	6

Twice the number of links gives the sum of the degrees of the vertices. (Or half the sum of the degrees of the vertices gives the number of links.)

B5

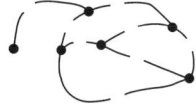

Look carefully at this diagram.
Each link connects two vertices so each link is counted twice when you add up the degrees of all the vertices.

C1

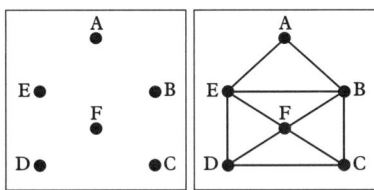

This is one route: C to E, E to B, B to C, C to D, D to B, B to A, A to E, E to D. Check your own routes, remember you cannot take your pencil off the paper or draw any line twice.
It does make a difference which dot you start with!
Did you try to find all the routes? There are two possible starting points and well over 50 possibilities.

C2 It is impossible to draw this without taking your pencil off the paper or without drawing any line twice!

C3 Rachel and Matthew's results are correct. Their report could perhaps have been a little clearer.

C4 Your own ideas here.

C5 The rule which works all the time is:
▲ Traceable networks never have more than two odd degree vertices. Check with your teacher if you're not sure.

D1 Here is the completed network for this park.

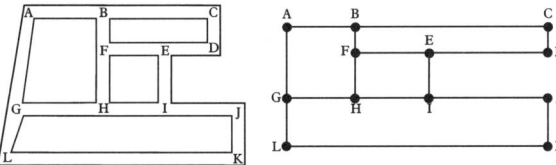

D2 What this question is really asking is 'Is the network traceable?'
There are six vertices of degree 3. Is it traceable?

D3 This network shows the path leaving at G.

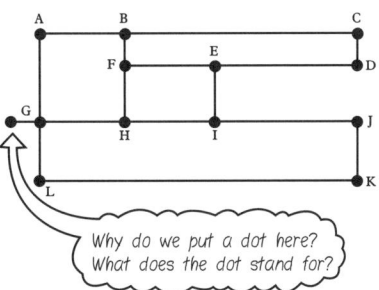

Why do we put a dot here?
What does the dot stand for?

D4
▲

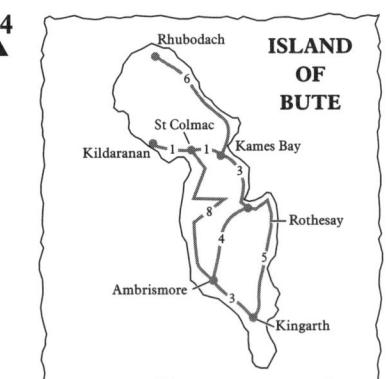

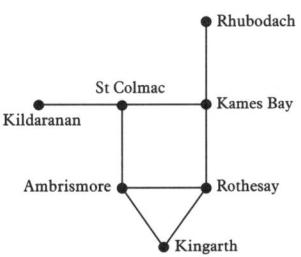

(a) Your own network. How could you check it against the map? One possibility is shown above.

(b) The only odd vertices are two of degree 1 and four of degree 3.
Is the network traceable?

(c) Here is one route which visits all the towns. Can you improve on it?
Rothesay – Kingarth – Ambrismore – St Colmac – Kildaranan – St Colmac – Kames Bay – Rhubodach – Kames Bay – Rothesay
A total of 34 miles.

D5

Here is a network for the shopping precinct.

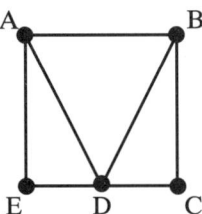

It has only two odd vertices (A and B). This means it is traceable. The only starting and finishing points (entrances and exits) possible are A and B. *Can you figure out why?*

D6

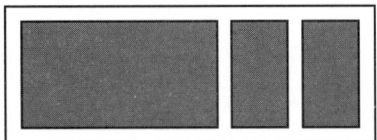

Here is a network for this map of the shopping precinct.

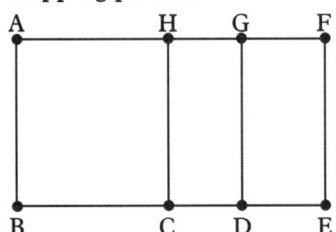

This network has four odd degree vertices (H, G, D and C).

This means it is not traceable. You will see some of the shops twice. This may not always be a bad thing for those shops! Probably the best places for an entrance and exit would be A and E or B and F. Why?

D7
▲

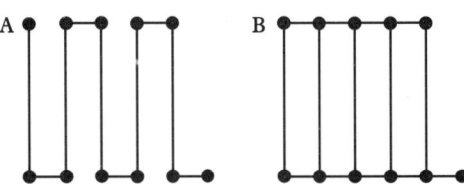

A is probably better for the company, because all the aisles are visited.
B is probably more convenient for the customer because they can go more or less directly to where they want.

D8

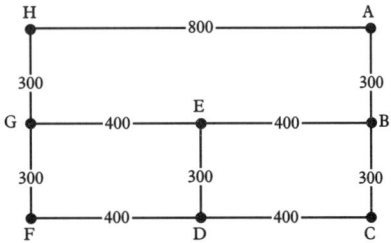

The network is not traceable. This means that Shenaz will need to retrace her steps. One possible route is A to B (300 m), B to C (300 m), C to D (400 m), D to F (400 m), F to G (300 m), G to E (400 m), E to D (300 m), D to C (400 m), C to B (300 m), B to E (400 m), E to G (400 m), G to H (300 m) and H to A (800 m). A total distance of 5000 metres. *You should be able to improve on this. Can you get as low as 4200 metres? (Your lettering may be slightly different to that used here.)*

D9

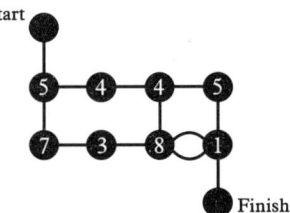

For this network the highest score is 42, which can be obtained in more than one way. This is one possibility.

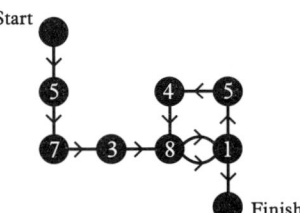

D10

Start

| 2 | 2 | 5 | 4 |
| 4 | 1 | 1 | 5 |
| 9 | 9 | 0 | 7 | Finish

For this plan the highest score is 48.
It can be obtained like this:

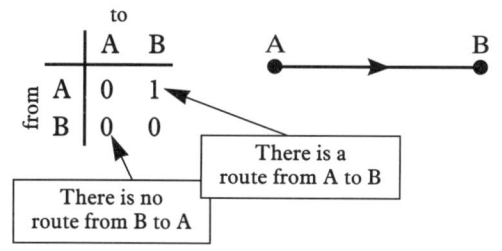

D11 Your own number mazes.

Challenges

- **Does this rule work?**
  For a network to be traceable and have
  the same starting and finishing point
  there must be no vertices of odd degree
  at all.
  (Can you figure out **why** the rule
  might work?)
- A way to take account of one-way
  routes in a link table is to read the
  table like this.

	to	
from		A	B
A	0	1
B	0	0

A ————————→ B

There is a
route from A to B

There is no
route from B to A

Vectors 1

A1

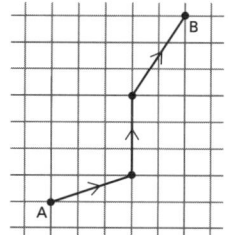

(a) The two other shifts needed to get
from A to B are:
$$\begin{bmatrix} 0 \\ 3 \end{bmatrix} \text{ and } \begin{bmatrix} 2 \\ 3 \end{bmatrix}.$$

(b) You finish at B.
(c) Yes, you always finish at B whatever
the order of the shifts.

A2

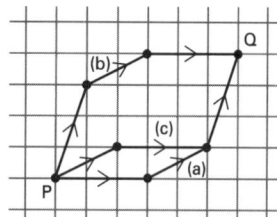

A3

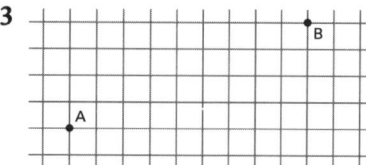

(a) These three shifts will get you from A
to B.
$$\begin{bmatrix} 4 \\ 1 \end{bmatrix}, \begin{bmatrix} 2 \\ 3 \end{bmatrix}, \begin{bmatrix} 3 \\ 0 \end{bmatrix}$$

(b) It does not matter in which order.

A4

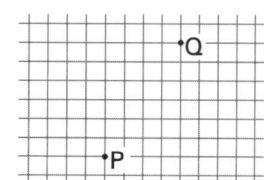

These three shifts will get you from P to Q.
$$\begin{bmatrix} 3 \\ 1 \end{bmatrix}, \begin{bmatrix} 1 \\ 3 \end{bmatrix}, \begin{bmatrix} 0 \\ 2 \end{bmatrix}$$

(The order does not matter.)

11

A5

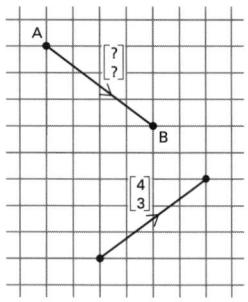

The shift needed to move from A to B is

$$\begin{bmatrix} 4 \\ -3 \end{bmatrix}.$$

B1

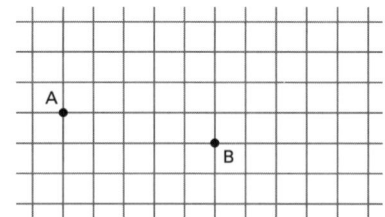

Line	Shift
(a)	$\begin{bmatrix} 1 \\ -3 \end{bmatrix}$
(b)	$\begin{bmatrix} -3 \\ -2 \end{bmatrix}$
(c)	$\begin{bmatrix} -2 \\ 4 \end{bmatrix}$
(d)	$\begin{bmatrix} 4 \\ -1 \end{bmatrix}$
(e)	$\begin{bmatrix} 2 \\ 2 \end{bmatrix}$
(f)	$\begin{bmatrix} -2 \\ -2 \end{bmatrix}$
(g)	$\begin{bmatrix} -4 \\ 0 \end{bmatrix}$
(h)	$\begin{bmatrix} 0 \\ -4 \end{bmatrix}$
(i)	$\begin{bmatrix} -5 \\ 3 \end{bmatrix}$

B2

(a) These shifts will get from A to B.

$$\begin{bmatrix} 3 \\ 2 \end{bmatrix} \text{ and } \begin{bmatrix} 2 \\ -3 \end{bmatrix}$$

(b) The order does not matter.

B3 These three shifts will get you from A to B.

$$\begin{bmatrix} -1 \\ -2 \end{bmatrix}, \begin{bmatrix} -1 \\ 1 \end{bmatrix}, \begin{bmatrix} 7 \\ 0 \end{bmatrix}$$

(The order does not matter.)

B4 Remember the inverse takes you back to the start.

(a) The inverse of $\begin{bmatrix} -3 \\ 4 \end{bmatrix}$ is $\begin{bmatrix} 3 \\ -4 \end{bmatrix}$.

(b) The inverse of $\begin{bmatrix} 3 \\ 3 \end{bmatrix}$ is $\begin{bmatrix} -3 \\ -3 \end{bmatrix}$.

(c) The inverse of $\begin{bmatrix} -1 \\ 3 \end{bmatrix}$ is $\begin{bmatrix} 1 \\ -3 \end{bmatrix}$.

(d) The inverse of $\begin{bmatrix} 0 \\ -2 \end{bmatrix}$ is $\begin{bmatrix} 0 \\ 2 \end{bmatrix}$.

(e) The inverse of $\begin{bmatrix} 3 \\ 0 \end{bmatrix}$ is $\begin{bmatrix} -3 \\ 0 \end{bmatrix}$.

B5 The secret message is:
DO YOU KNOW THE WORD VECTOR

B6

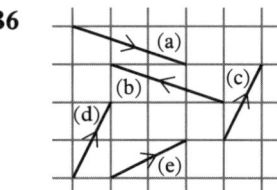

Here are the column vectors:

(a) $\begin{bmatrix} 3 \\ -1 \end{bmatrix}$ (b) $\begin{bmatrix} -3 \\ 1 \end{bmatrix}$ (c) $\begin{bmatrix} 1 \\ 2 \end{bmatrix}$ (d) $\begin{bmatrix} 1 \\ 2 \end{bmatrix}$ (e) $\begin{bmatrix} 2 \\ 1 \end{bmatrix}$

B7 (c) and (d) are equal vectors.

They are both $\begin{bmatrix} 1 \\ 2 \end{bmatrix}$.

C1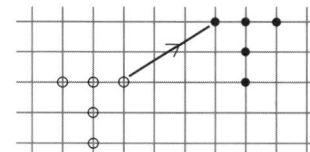

(a) Karl was wrong – he wasn't looking at similar points.

(b) The answer he should have got
was $\begin{bmatrix} 5 \\ 2 \end{bmatrix}$.

C2 The hollow circles have been translated by $\begin{bmatrix} 5 \\ -1 \end{bmatrix}$ to give the filled circles.

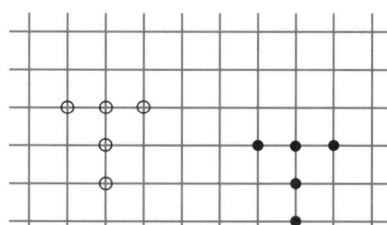

C3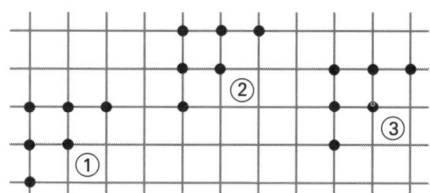

(a) To move from ① to ② the translation
vector is $\begin{bmatrix} 4 \\ 2 \end{bmatrix}$.

(b) To move from ② to ③ the translation
vector is $\begin{bmatrix} 4 \\ -1 \end{bmatrix}$.

(c) To move from ① to ③ the translation
vector is $\begin{bmatrix} 8 \\ 1 \end{bmatrix}$.

(d) To move from ③ to ② the translation
vector is $\begin{bmatrix} -4 \\ 0 \end{bmatrix}$.

(e) To move from ③ to ① the translation
vector is $\begin{bmatrix} -8 \\ -1 \end{bmatrix}$.

C4 The dotted triangle has been translated into the solid triangle by the column vector $\begin{bmatrix} -2 \\ -1 \end{bmatrix}$.

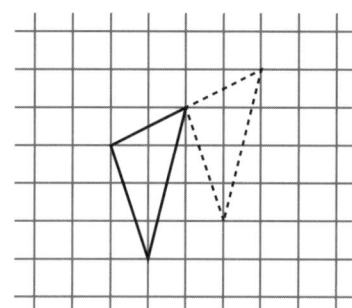

C5

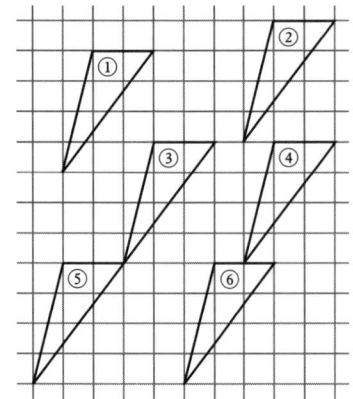

	Translation from	Column vector
(a)	① to ②	$\begin{bmatrix} 6 \\ 1 \end{bmatrix}$
(b)	② to ①	$\begin{bmatrix} ^-6 \\ ^-1 \end{bmatrix}$
(c)	③ to ④	$\begin{bmatrix} 4 \\ 0 \end{bmatrix}$
(d)	④ to ③	$\begin{bmatrix} ^-4 \\ 0 \end{bmatrix}$
(e)	③ to ⑤	$\begin{bmatrix} ^-3 \\ ^-4 \end{bmatrix}$
(f)	⑤ to ③	$\begin{bmatrix} 3 \\ 4 \end{bmatrix}$
(g)	④ to ⑥	$\begin{bmatrix} ^-2 \\ ^-4 \end{bmatrix}$
(h)	⑥ to ④	$\begin{bmatrix} 2 \\ 4 \end{bmatrix}$
(i)	② to ⑥	$\begin{bmatrix} ^-2 \\ ^-8 \end{bmatrix}$
(j)	⑥ to ②	$\begin{bmatrix} 2 \\ 8 \end{bmatrix}$
(k)	⑤ to ④	$\begin{bmatrix} 7 \\ 4 \end{bmatrix}$
(l)	④ to ⑤	$\begin{bmatrix} ^-7 \\ ^-4 \end{bmatrix}$

D1   Show your pattern to your teacher.

E1

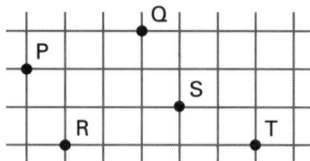

(a) $\overrightarrow{PQ} = \begin{bmatrix} 3 \\ 1 \end{bmatrix}$ (b) $\overrightarrow{QP} = \begin{bmatrix} ^-3 \\ ^-1 \end{bmatrix}$ (c) $\overrightarrow{RS} = \begin{bmatrix} 3 \\ 1 \end{bmatrix}$

(d) $\overrightarrow{SR} = \begin{bmatrix} ^-3 \\ ^-1 \end{bmatrix}$ (e) $\overrightarrow{RT} = \begin{bmatrix} 5 \\ 0 \end{bmatrix}$ (f) $\overrightarrow{TR} = \begin{bmatrix} ^-5 \\ 0 \end{bmatrix}$

(g) $\overrightarrow{TQ} = \begin{bmatrix} ^-3 \\ 3 \end{bmatrix}$ (h) $\overrightarrow{QT} = \begin{bmatrix} 3 \\ ^-3 \end{bmatrix}$

(i)   Vector $\overrightarrow{SR}$ is equal to $\overrightarrow{QP}$.

E2

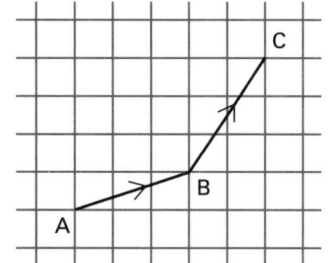

In the diagram $\begin{bmatrix} 5 \\ 4 \end{bmatrix}$ is the vector $\overrightarrow{AC}$.

E3

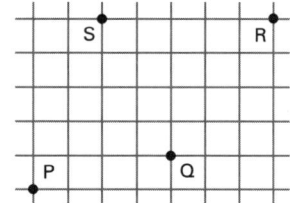

(a) $\overrightarrow{PQ} = \begin{bmatrix} 4 \\ 1 \end{bmatrix}$ (b) $\overrightarrow{QR} = \begin{bmatrix} 3 \\ 4 \end{bmatrix}$

(c) $\overrightarrow{PQ} + \overrightarrow{QR} = \begin{bmatrix} 7 \\ 5 \end{bmatrix}$ (d) $\overrightarrow{PS} = \begin{bmatrix} 2 \\ 5 \end{bmatrix}$

(e) $\overrightarrow{SR} = \begin{bmatrix} 5 \\ 0 \end{bmatrix}$ (f) $\overrightarrow{PS} + \overrightarrow{SR} = \begin{bmatrix} 7 \\ 5 \end{bmatrix}$

(g) $\overrightarrow{PR} = \begin{bmatrix} 7 \\ 5 \end{bmatrix}$

E4

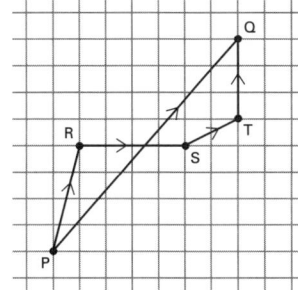

(a) $\overrightarrow{PQ} = \begin{bmatrix} 7 \\ 8 \end{bmatrix}$

(b) $\begin{bmatrix} 1 \\ 4 \end{bmatrix} + \begin{bmatrix} 4 \\ 0 \end{bmatrix} + \begin{bmatrix} 2 \\ 1 \end{bmatrix} + \begin{bmatrix} 0 \\ 3 \end{bmatrix} = \begin{bmatrix} 7 \\ 8 \end{bmatrix}$

E5
(a) $7 + {}^-4 = 3$ (b) ${}^-1 + {}^-2 = {}^-3$
(c) $10 + {}^-4 = 6$ (d) $3 + {}^-5 = {}^-2$
(e) ${}^-2 + {}^-3 = {}^-5$ (f) ${}^-2 + 4 = 2$
(g) ${}^-1 + {}^-5 = {}^-6$ (h) $3 + {}^-3 = 0$
(i) $2 + {}^-7 = {}^-5$

E6

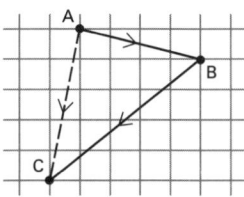

(a) $\overrightarrow{AB} = \begin{bmatrix} 4 \\ {}^-1 \end{bmatrix}$ (b) $\overrightarrow{BC} = \begin{bmatrix} {}^-5 \\ {}^-4 \end{bmatrix}$

(c) $\overrightarrow{AB} + \overrightarrow{BC} = \begin{bmatrix} {}^-1 \\ {}^-5 \end{bmatrix}$

E7

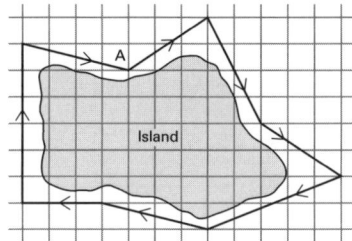

(a) Here are the column vectors for each stage of the journey.

$\begin{bmatrix} 3 \\ 2 \end{bmatrix} \begin{bmatrix} 2 \\ {}^-4 \end{bmatrix} \begin{bmatrix} 3 \\ {}^-2 \end{bmatrix} \begin{bmatrix} {}^-5 \\ {}^-2 \end{bmatrix} \begin{bmatrix} {}^-4 \\ 1 \end{bmatrix} \begin{bmatrix} {}^-3 \\ 0 \end{bmatrix} \begin{bmatrix} 0 \\ 6 \end{bmatrix} \begin{bmatrix} 4 \\ {}^-1 \end{bmatrix}$

(b) These eight vectors added together give $\begin{bmatrix} 0 \\ 0 \end{bmatrix}$.

(c) The journey starts and finishes at the same point. This means that the overall effect is $\begin{bmatrix} 0 \\ 0 \end{bmatrix}$.

Puzzle

The pattern is made up from a block of 12 letters.
This block is translated with the vector
$\begin{bmatrix} 4 \\ 2 \end{bmatrix}$ or $\begin{bmatrix} 4 \\ {}^-1 \end{bmatrix}$ or $\begin{bmatrix} {}^-4 \\ {}^-2 \end{bmatrix}$ or $\begin{bmatrix} {}^-4 \\ 1 \end{bmatrix}$.

Vectors 2

A1

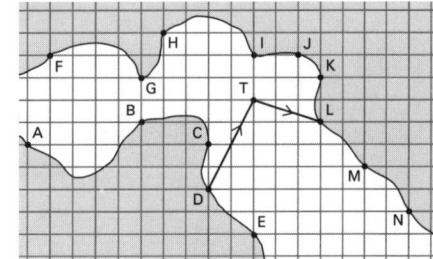

A boat setting out from B which makes the journey $\begin{bmatrix} 2 \\ 4 \end{bmatrix}$ and $\begin{bmatrix} 3 \\ {}^-1 \end{bmatrix}$ finishes at I.

A2 A boat which travels $\begin{bmatrix} 2 \\ 4 \end{bmatrix}$ then $\begin{bmatrix} 3 \\ {}^-1 \end{bmatrix}$ and sets out from:
(a) A will finish at G
(b) C will finish at K
(c) E will finish at M.

A3 The column vector for $\overrightarrow{DL}$ is $\begin{bmatrix} 5 \\ 3 \end{bmatrix}$.

A4 The column vector for:

(a) $\overrightarrow{AG}$ is $\begin{bmatrix} 5 \\ 3 \end{bmatrix}$ (b) $\overrightarrow{BI}$ is $\begin{bmatrix} 5 \\ 3 \end{bmatrix}$

(c) $\overrightarrow{CK}$ is $\begin{bmatrix} 5 \\ 3 \end{bmatrix}$ (d) $\overrightarrow{EM}$ is $\begin{bmatrix} 5 \\ 3 \end{bmatrix}$.

The vector $\overrightarrow{DL}$ is called the **sum** of $\overrightarrow{DT}$ and $\overrightarrow{TL}$.
If you add the column vectors of $\overrightarrow{DT}$ and $\overrightarrow{TL}$,
like this, you get the column vector of $\overrightarrow{DL}$.

$\overrightarrow{DT} + \overrightarrow{TL} = \overrightarrow{DL}$ Add the top numbers.
$\begin{bmatrix} 2 \\ 4 \end{bmatrix} + \begin{bmatrix} 3 \\ -1 \end{bmatrix} = \begin{bmatrix} 5 \\ 3 \end{bmatrix}$ $2 + 3 = 5$
    Add the bottom numbers
    $4 + {}^-1 = 3$

A5 If you get any of these wrong, check back using the map of the river estuary.

(a) $\overrightarrow{AG} + \overrightarrow{GI} = \overrightarrow{AI}$
$\begin{bmatrix} 5 \\ 3 \end{bmatrix} + \begin{bmatrix} 5 \\ 1 \end{bmatrix} = \begin{bmatrix} 10 \\ 4 \end{bmatrix}$

(b) $\overrightarrow{HK} + \overrightarrow{KE} = \overrightarrow{HE}$
$\begin{bmatrix} 7 \\ -2 \end{bmatrix} + \begin{bmatrix} -3 \\ -7 \end{bmatrix} = \begin{bmatrix} 4 \\ -9 \end{bmatrix}$

(c) $\overrightarrow{NC} + \overrightarrow{CD} = \overrightarrow{ND}$
$\begin{bmatrix} -9 \\ 3 \end{bmatrix} + \begin{bmatrix} 0 \\ -2 \end{bmatrix} = \begin{bmatrix} -9 \\ 1 \end{bmatrix}$

(d) $\overrightarrow{JI} + \overrightarrow{IB} = \overrightarrow{JB}$
$\begin{bmatrix} -2 \\ 0 \end{bmatrix} + \begin{bmatrix} -5 \\ -3 \end{bmatrix} = \begin{bmatrix} -7 \\ -3 \end{bmatrix}$

A6 If $\underset{\sim}{a} = \begin{bmatrix} 4 \\ 1 \end{bmatrix}$ and $\underset{\sim}{b} = \begin{bmatrix} 2 \\ 3 \end{bmatrix}$ then

$\underset{\sim}{a} + \underset{\sim}{b} = \begin{bmatrix} 4 \\ 1 \end{bmatrix} + \begin{bmatrix} 2 \\ 3 \end{bmatrix} = \begin{bmatrix} 6 \\ 4 \end{bmatrix}$.

A7

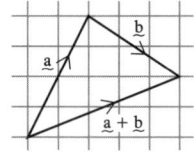

A8 (a) $\underset{\sim}{a} + \underset{\sim}{b} = \begin{bmatrix} -3 \\ -2 \end{bmatrix}$

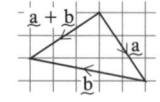

(b) $\underset{\sim}{b} + \underset{\sim}{a} = \begin{bmatrix} -3 \\ -2 \end{bmatrix}$

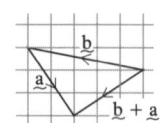

(c) $\underset{\sim}{a} + \underset{\sim}{c} = \begin{bmatrix} 1 \\ -7 \end{bmatrix}$

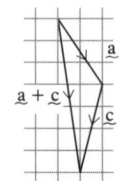

(d) $\underset{\sim}{c} + \underset{\sim}{a} = \begin{bmatrix} 1 \\ -7 \end{bmatrix}$

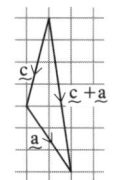

A9

	Vector	Inverse
(a)	$\begin{bmatrix} 2 \\ 5 \end{bmatrix}$	$\begin{bmatrix} -2 \\ -5 \end{bmatrix}$
(b)	$\begin{bmatrix} -6 \\ -1 \end{bmatrix}$	$\begin{bmatrix} 6 \\ 1 \end{bmatrix}$
(c)	$\begin{bmatrix} -3 \\ 2 \end{bmatrix}$	$\begin{bmatrix} 3 \\ -2 \end{bmatrix}$
(d)	$\begin{bmatrix} 4 \\ -7 \end{bmatrix}$	$\begin{bmatrix} -4 \\ 7 \end{bmatrix}$
(e)	$\begin{bmatrix} 0 \\ 2 \end{bmatrix}$	$\begin{bmatrix} 0 \\ -2 \end{bmatrix}$
(f)	$\begin{bmatrix} -4 \\ 0 \end{bmatrix}$	$\begin{bmatrix} 4 \\ 0 \end{bmatrix}$

A10 (a) If $\underset{\sim}{a} = \begin{bmatrix} 2 \\ -5 \end{bmatrix}$, then $-\underset{\sim}{a} = \begin{bmatrix} -2 \\ 5 \end{bmatrix}$.

(b) $\underset{\sim}{a} + {}^-\underset{\sim}{a} = \begin{bmatrix} 2 \\ -5 \end{bmatrix} + \begin{bmatrix} -2 \\ 5 \end{bmatrix} = \begin{bmatrix} 0 \\ 0 \end{bmatrix}$

(c) Whatever values you choose

$\underset{\sim}{a} + {}^-\underset{\sim}{a} = \begin{bmatrix} 0 \\ 0 \end{bmatrix}$.

A11 If $\underset{\sim}{a} = \begin{bmatrix} 4 \\ -3 \end{bmatrix}$, $\underset{\sim}{b} = \begin{bmatrix} 2 \\ 5 \end{bmatrix}$ and $\underset{\sim}{c} = \begin{bmatrix} 6 \\ 2 \end{bmatrix}$ then:

(a) $-\underset{\sim}{a} = \begin{bmatrix} 4 \\ -3 \end{bmatrix}$　　(b) $-\underset{\sim}{b} = \begin{bmatrix} -2 \\ -5 \end{bmatrix}$

(c) $-\underset{\sim}{c} = \begin{bmatrix} -6 \\ -2 \end{bmatrix}$　　(d) $-\underset{\sim}{a} + \underset{\sim}{b} = \begin{bmatrix} -2 \\ 8 \end{bmatrix}$

(e) $\underset{\sim}{a} + -\underset{\sim}{b} = \begin{bmatrix} 2 \\ -8 \end{bmatrix}$ (f) $-\underset{\sim}{a} + -\underset{\sim}{b} = \begin{bmatrix} -6 \\ -2 \end{bmatrix}$

(g) $\underset{\sim}{a} + \underset{\sim}{b} + -\underset{\sim}{c} = \begin{bmatrix} 0 \\ 0 \end{bmatrix}$.

B1

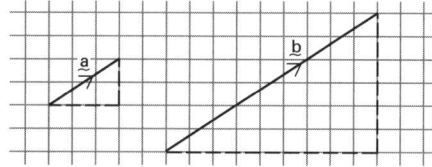

(a) $\underset{\sim}{a} = \begin{bmatrix} 3 \\ 2 \end{bmatrix}$　　(b) $\underset{\sim}{b} = \begin{bmatrix} 9 \\ 6 \end{bmatrix}$

(What do you notice about these two column vectors?)

B2 If $\underset{\sim}{p} = \begin{bmatrix} 4 \\ -2 \end{bmatrix}$, then: (a) $2\underset{\sim}{p} = \begin{bmatrix} 8 \\ -4 \end{bmatrix}$

(b) $5\underset{\sim}{p} = \begin{bmatrix} 20 \\ -10 \end{bmatrix}$ and (c) $\frac{1}{2}\underset{\sim}{p} = \begin{bmatrix} 2 \\ -1 \end{bmatrix}$.

B3 If $\underset{\sim}{a} = \begin{bmatrix} 3 \\ 2 \end{bmatrix}$ and $\underset{\sim}{b} = \begin{bmatrix} -2 \\ 5 \end{bmatrix}$, then:

(a) $4\underset{\sim}{a} = \begin{bmatrix} 12 \\ 8 \end{bmatrix}$　　(b) $3\underset{\sim}{b} = \begin{bmatrix} -6 \\ 15 \end{bmatrix}$

(c) $4\underset{\sim}{a} + 3\underset{\sim}{b} = \begin{bmatrix} 6 \\ 23 \end{bmatrix}$ (d) $2\underset{\sim}{a} + 5\underset{\sim}{b} = \begin{bmatrix} -4 \\ 29 \end{bmatrix}$.

B4 (a) (b)

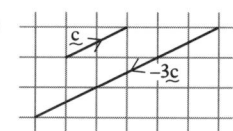

(c) $-3\underset{\sim}{c}$ is 3 times as long as $\underset{\sim}{c}$ and points in the opposite direction to $\underset{\sim}{c}$.

B5 (a) (b) (c)

$-2\underset{\sim}{d} = \begin{bmatrix} 4 \\ -6 \end{bmatrix}$

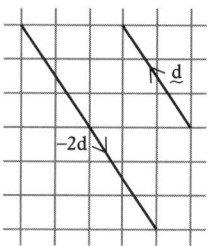

B6 If $\underset{\sim}{e} = \begin{bmatrix} -4 \\ -1 \end{bmatrix}$, then $-5\underset{\sim}{e} = \begin{bmatrix} 20 \\ 5 \end{bmatrix}$.

B7

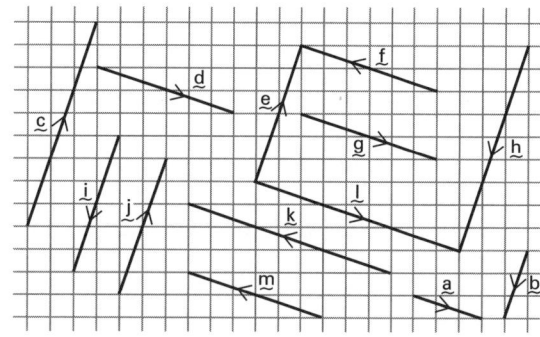

(a) $\underset{\sim}{d}$ and $\underset{\sim}{g}$ are both equal to $2\underset{\sim}{a}$.
(b) $\underset{\sim}{i}$ is equal to $2\underset{\sim}{b}$.
(c) $\underset{\sim}{f}$ and $\underset{\sim}{m}$ are both equal to $-2\underset{\sim}{a}$.
(d) $\underset{\sim}{e}$ and $\underset{\sim}{j}$ are both equal to $-2\underset{\sim}{b}$.
(e) $\underset{\sim}{l}$ is equal to $3\underset{\sim}{a}$.
(f) $\underset{\sim}{h}$ is equal to $-3\underset{\sim}{b}$.
(g) $\underset{\sim}{k}$ is equal to $-3\underset{\sim}{a}$.
(h) $\underset{\sim}{c}$ is equal to $-3\underset{\sim}{b}$.

C1 $\underset{\sim}{r} = 2\underset{\sim}{p} + 5\underset{\sim}{q}$
This diagram may help.

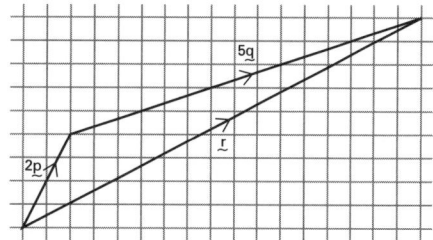

C2 $\underset{\sim}{c} = 6\underset{\sim}{a} + 2\underset{\sim}{b}$
Show your triangle to your teacher.

C3 $\underset{\sim}{w} = 3\underset{\sim}{u} + \underset{\sim}{v}$
Show your triangle to your teacher.

C4 $\underset{\sim}{c} = 23\underset{\sim}{a} + 10\underset{\sim}{b}$

C5 $\underset{\sim}{r} = 5\underset{\sim}{p} + 2\underset{\sim}{q}$

C6 $\underset{\sim}{w} = 3\underset{\sim}{u} + 4\underset{\sim}{v}$

C7 $\underset{\sim}{b}$ is parallel to $\underset{\sim}{a}$, so the result of any linear sum of $\underset{\sim}{a}$ and $\underset{\sim}{b}$ is also parallel to $\underset{\sim}{a}$ and $\underset{\sim}{b}$. Since $\underset{\sim}{c}$ is not parallel to $\underset{\sim}{a}$ it cannot be obtained using only $\underset{\sim}{a}$ and $\underset{\sim}{b}$.

*If you want to be able to get **any** vector as a linear sum you must use two vectors which are not parallel.*

Movement

Mirrors and reflection 1

A1–A5 See your teacher if you could not do any of these.
Before you do, work together to try to do them!

B1
(b) (c) (d)
(e) (f) (g)
(h) (i) (j)
(k) cannot be done. (l)

B2 (a) DID DICK HIDE BOX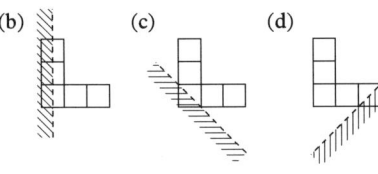

(b) DEBBIE HID BOX

(c) DICK KICKED DEBBIE

B3
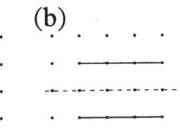
∩∨   ∨I∩
or   or
∪∧   ∧I∪

CH∩∨E
or
CH∪∧E

H∩∩∩E∩
or
H∪∪∪E∪

B4 All these letters 'go wrong' when you put a mirror along the dotted line: A, F, G, J, L, M, N, P, Q, R, S, T, U, V, W, Y, Z. You may also have noticed that the top and bottom of the B on the worksheet are slightly different!

~~ABCDEFGHIJKLMNOPQRSTUVWXYZ~~

B5 Your own name in mirror code

B6 You should have checked these for yourselves, but just in case:
(a) $31 - 13 = 18$ (b) $31 - 18 = 13$
(c) $308 + 10 = 318$ (d) $0 \times 8 = 0$.

C1 Your own work

C2 (a)   (b)   (c)
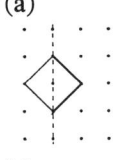

C3 (a)   (b)   (c)

(e)   (f)   (g)
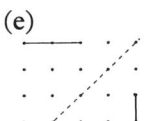 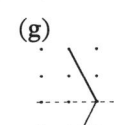

C4
(a)   (b)
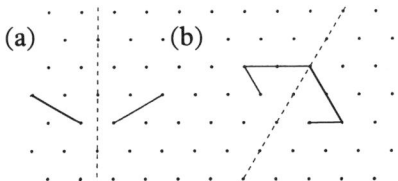

C5

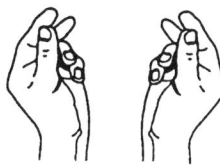

The two shapes are as exactly the same as a person's two hands (or the two front doors of a car). They seem the same, but how are they different?

C6　Your own work

D1　The cat's face does not have reflection symmetry.

D2　These signs have reflection symmetry.

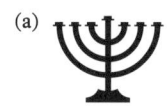 (a)　(b) 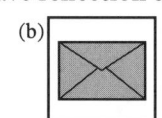　(f)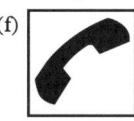

These do not have reflection symmetry.

(c) 　(d) 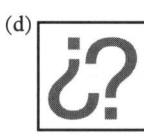　(e)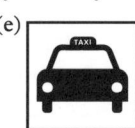

Check your answers again if you are not sure.

D3　The answers are at the back of the booklet.

D4　These signs have reflection symmetry.

(b) 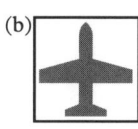　(e)

These do not have reflection symmetry.

(a)  (c)  (d)  (f)

Check your answers again if you are not sure.

D5　The answers are at the back of the booklet.

D6　The dotted line (where a mirror would fit) is called a line of reflection symmetry.

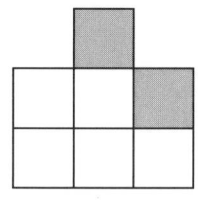

This does not have reflection symmetry.　*But this shape does. The dotted line shows it.*

D7　The dotted line (where a mirror would fit) is called a line of reflection symmetry (or sometimes just a line of symmetry).

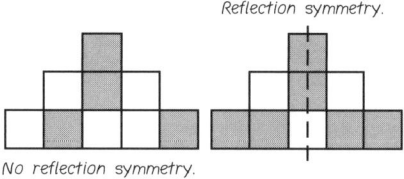

Reflection symmetry.

No reflection symmetry.

D8　These pictures have reflection symmetry. The dotted lines show the lines of reflection symmetry. Test them with a mirror if you are not sure.

D9　The planes of symmetry for this shape:

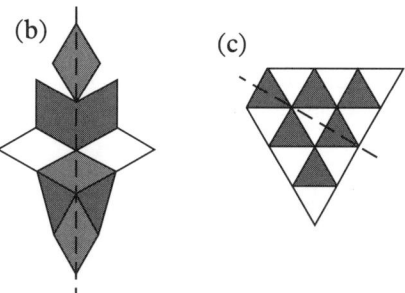

are:

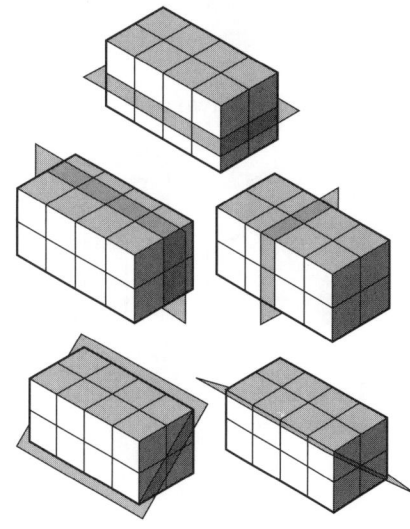

19

D10

Shoes and scissors do not have a plane of symmetry.

A cup and saucer or a pen may have a plane of symmetry. (It depends on their exact shape. For example, a pen may have a plane of symmetry if you ignore its screw cap or the writing on its side, etc.)

D11 Some possible shapes are shown here.
▲ There are other possibilities, so you may have something different. If you do, show it to your teacher.

(a) A shape with no planes of symmetry.

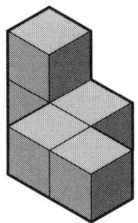

(b) A shape with one plane of symmetry.

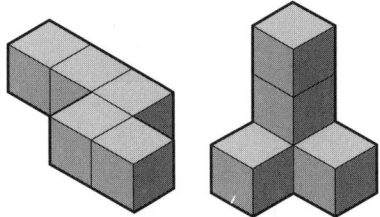

(c) A shape with two planes of symmetry.

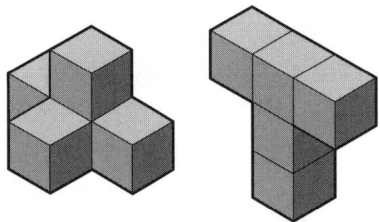

(d) There are no shapes with three planes of symmetry.

D12 A cube has nine planes of symmetry. They are shown below.

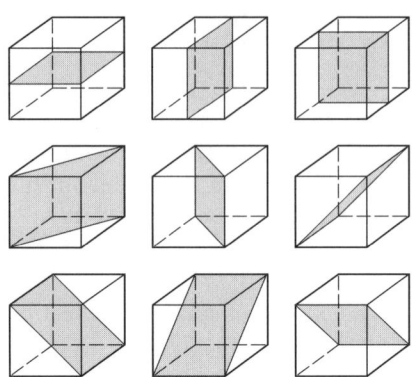

Challenge

Here is a drawing of Claire's cube.
It is made from 6 white and 2 red cubes and has three planes of symmetry.

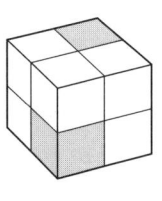

D13
▲ (a) 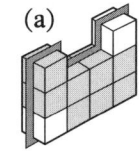 (b) 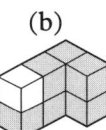 (c)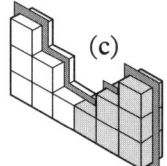

(a) This shape has one plane of symmetry.
(b) This shape has no planes of symmetry.
(c) this shape has one plane of symmetry.

D14 (a) These solids have no planes of symmetry.

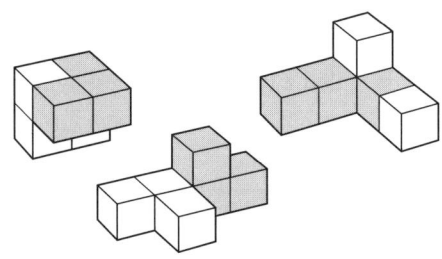

(b) These solids have two planes of symmetry.

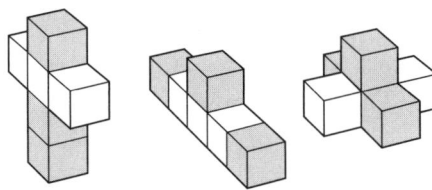

Reflection 2

A1 Here is what is wrong when you look at each complete picture when you put a mirror on the dotted lines.

(a) The numbers are back to front in the reflected part.
Have you ever tried to read a newspaper in a mirror?

(b) There are two clips.

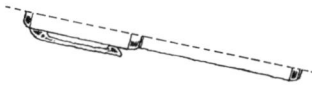

(c) There are two eyes on the same side.

(d) There are two thumbs and three fingers.

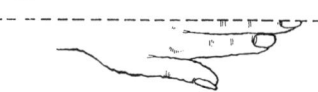

A2

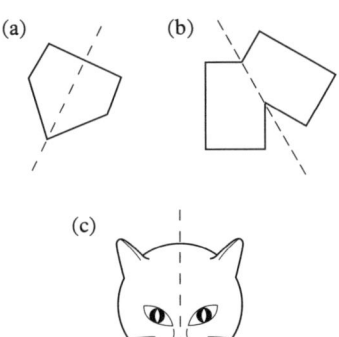

A3

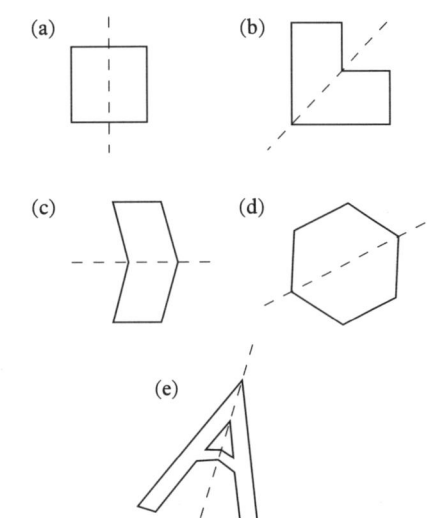

If you drew the A wrong, make sure you know how to do it right now – a lot of people get this wrong first time.

B1 Your rhombus should be congruent to this one here.
Remember, congruent means exactly the same shape as. How can you tell if two shapes are congruent?

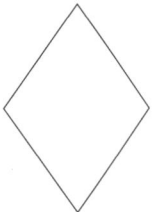

B2 When you fold a piece of paper into four ('twice-folded' paper):

(a) cut it like this 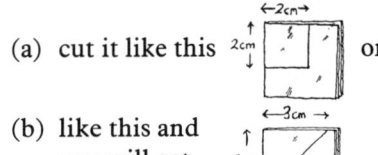 or

(b) like this and you will get a square.

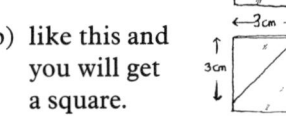

B3

When Sarah cuts the above shape and opens it out she will get this shape (Shape C).

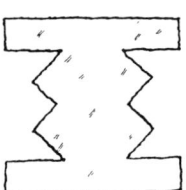

B4

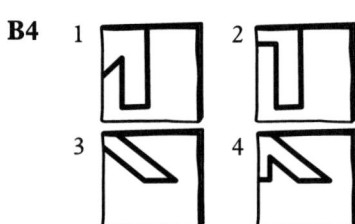

2 makes an H, 3 makes an X.

B5

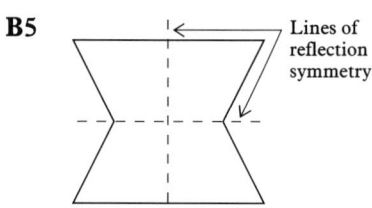

Lines of reflection symmetry

B6

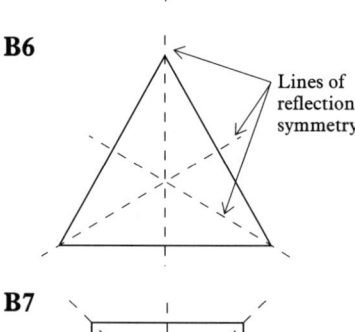

Lines of reflection symmetry

B7

B8

The parallelogram has no lines of symmetry.

B9 (a) F is the first letter with no line of symmetry.

   (b) H is the first letter with 2 lines of symmetry.

   (c) Only the letters with lines of symmetry are shown here.
   The others do not have reflection symmetry.

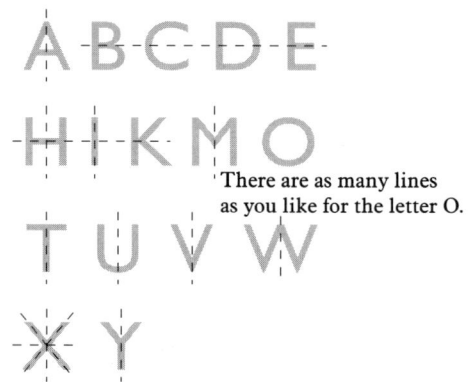

There are as many lines as you like for the letter O.

(d)

Number of lines of reflection symmetry	Letters
0	F G J L N P Q R S Z
1	A B C D E K M T U V W Y
2	H I
3	
4	X
more than 4	O

22

(a)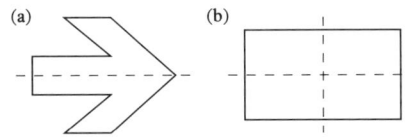

1 line of symmetry

(b)

2 lines of symmetry

(c)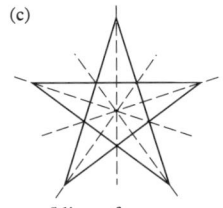

5 lines of symmetry

(d)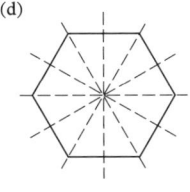

6 lines of symmetry

(e)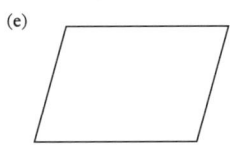

None

(f)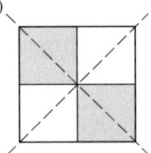

2 lines of symmetry

C1

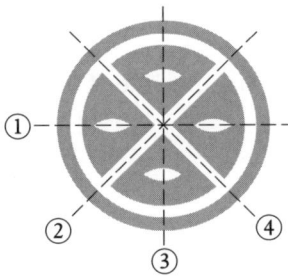

①and③are lines of symmetry.

C2 (a) ①and③are lines of symmetry.

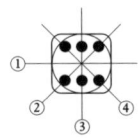

(b) ②is a line of symmetry.

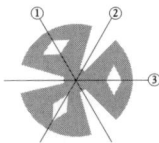

(c) Neither is a line of symmetry.

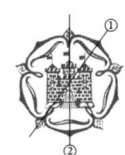

(d) ①,③and⑤are lines of symmetry.

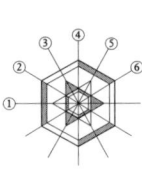

(e) ①and③are lines of symmetry.

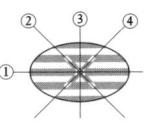

(f) Neither is a line of symmetry.

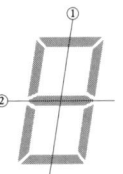

(g) Neither is a line of symmetry.

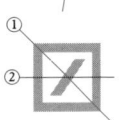

D1 (a) You can see 3 coins.

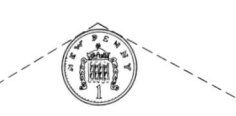

(b) You can see 4 coins.

(c) You can see 5 coins.

(d) You can see 6 coins.

D2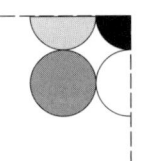

When you put a mirror along each dotted line you can see:
(a) 2 grey spots   (b) 1 black spot
(c) 2 white spots   (d) 4 red spots.

Here are the completed patterns for **D3** to **D8**.

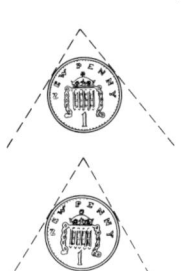

D3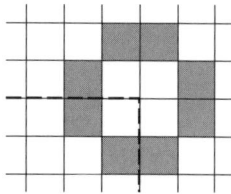

D4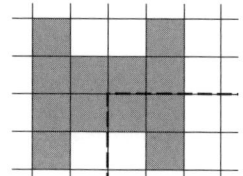

D5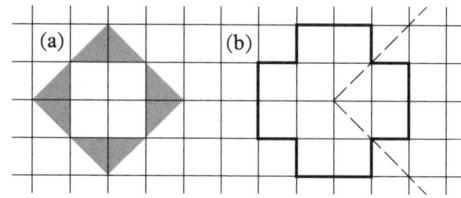

D6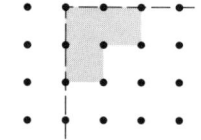

D7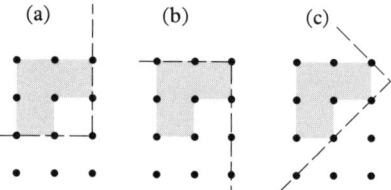

D8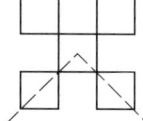

E1 There are many different ways, for example:

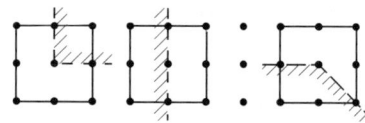

E2 There are many different answers, for example:

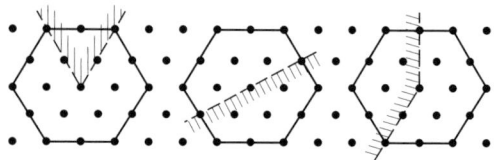

E3 There are many different ways, for example:

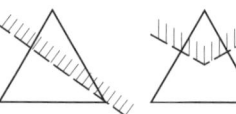 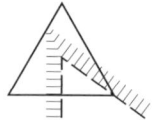

F1 (a) This pattern has two lines of reflection symmetry.

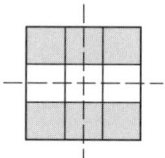

(b) All these shapes have one line of symmetry, and are made by swapping two tiles.

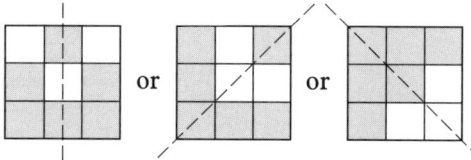

F2 (a) This shape has one line of symmetry.

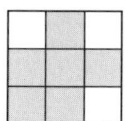

(b) Here is a pattern which can be made by swapping two tiles in the shape in (a). It has two lines of symmetry. You may have a different pattern. Check it with a mirror.

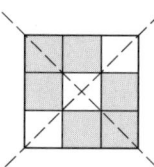

F3 These patterns have no lines of symmetry. You probably found some others.

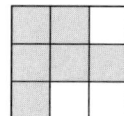 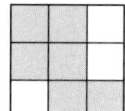 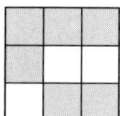

F4 It is not possible to make a square with the 9 tiles which has four lines of reflection symmetry.

F5

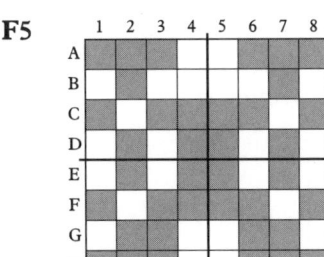

If the tiles B3 and B6 are changed from white to red the tiled floor will have two lines of reflection symmetry.

F6

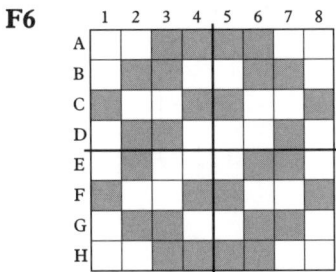

If the tiles D6 and E3 are changed from white to red the tiled floor will have two lines of reflection symmetry.

F7

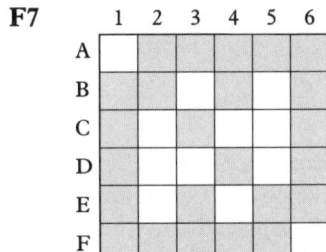

If the tiles C5 and D2 are changed from white to red the tiled floor will have two lines of reflection symmetry.

Reflection puzzles

A1

JABBERWOCKY
'Twas brillig, and the slithy toves
Did gyre and gimble in the wabe:
All mimsy were the the borogoves,
And the mome raths outgrabe.

means

JABBERWOCKY
'Twas brillig, and the slithy toves
Did gyre and gimble in the wabe:
All mimsy were the borogoves,
And the mome raths outgrabe.

A2   I HIT THAT TAXI WITH A TOMATO.

I HIT THAT TAXI WITH A TOMATO.

A3   I SAY TIMOTHY, WHAT A WAY TO HIT THAT HAT!

'THAT' should be 'TAHT'.
The 'S' in 'SAY' is in 'ordinary' writing.

A4   Check your own answer with a mirror.

A5   # TUO YAW

You might see this on a glass door seen from the outside or in a mirror facing an exit. It says 'WAY OUT'.

A6   The word 'fire' is printed in mirror writing so that it can be seen the right way round in a car mirror.

A7   AMBULANCE

The 'N' in this sign for 'ambulance' is not in mirror writing.

B1

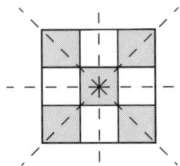

(a) This pattern has four lines of reflection symmetry.

(b) By swapping one black and one white tile you can make a new pattern which has no lines of reflection symmetry. Here is one way. You may have found some others – don't forget to check them.

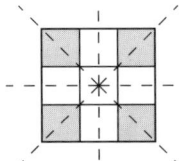

B2

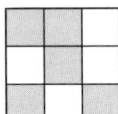

By changing the centre tile in pattern A (the top one in B1) you get another pattern which has four lines of symmetry.

B3 This pattern has two lines of reflection symmetry.

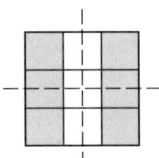

B4 This has four lines of reflection symmetry.

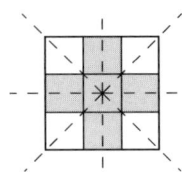

B5 (a) This pattern, made from one white and eight coloured tiles, has four lines of reflection symmetry.

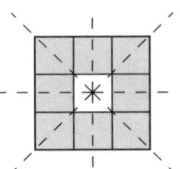

(b) These patterns, made from one white and eight coloured tiles all have one line of reflection symmetry. You may have found some different ones yourselves.

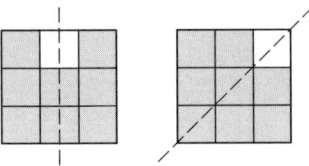

B6 There is always at least one line of symmetry in a square pattern made from eight coloured and one white tile.

B7

(a) If a coloured tile is placed at X the pattern will have no lines of symmetry.

(b) If a white tile is placed at X the pattern will have no lines of symmetry.

B8

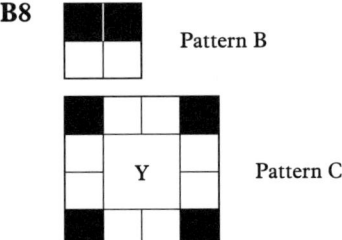

Pattern B

Pattern C

(a) If pattern B is placed at Y the new pattern will have 1 line of symmetry.

(b) If pattern B is turned through 90° and placed at Y the new pattern will have 1 line of symmetry.

C1

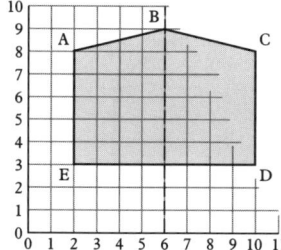

The dotted line is a line of reflection symmetry.

(a) C is at (10, 8).　　(b) D is at (10, 3).

How did you work these out without drawing?

C2

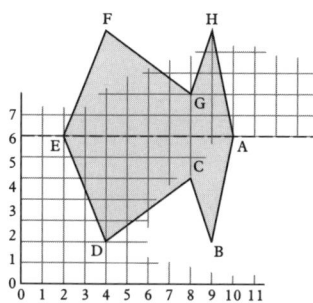

(a) B is at (9, 2).　　(b) F is at (4, 12).

C3

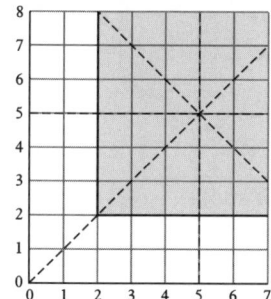

(a) One corner of the square is at (2, 2); the other three are at (8, 2), (2, 8) and (8, 8).

(b) The other spots are at (6, 2), (4, 8), (6, 8), (2, 4), (2, 6), (8, 4) and (8, 6).

C4 This shape when completed has two lines of reflection symmetry.

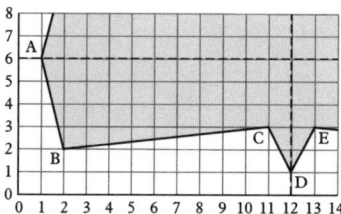

(a) The completed shape has 12 corners.

(b) These are the co-ordinates of the other seven corners.
(22, 2), (23, 6), (22, 10), (13, 9), (12, 11), (11, 9), (2, 10).

(c) If there is a spot at (17, 3) there need to be spots at (17, 9), (7, 3) and (7, 9) for the shape still to have two lines of reflection symmetry.

C5 The dotted line is a line of reflection symmetry.

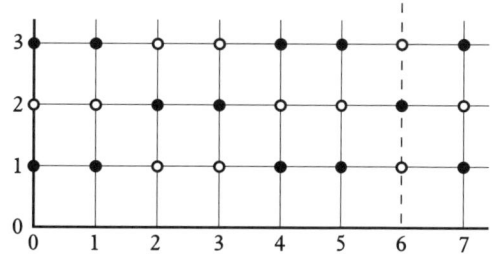

(a) ○ at (12, 2)　　(b) ○ at (10, 3)
(c) ○ at (9, 1)　　(d) ● at (11, 3)

You can check these yourselves by using a mirror.

D1

The clock shows 5 to 4.
Her brother will be ready at 4 o'clock.

D2

He had turned right.

D3 The driver would see this in the rear view mirror:
MOAT TAXIS 881818

D4 The clock was fast by about 25 minutes.

D5 The traffic is driving on the right, drivers are sitting on the left, the Co-op sign is reversed, and so on. *Use a mirror to check the others.*

D6

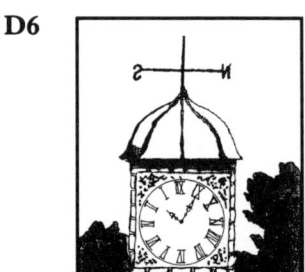

(a) The time is 1:55 p.m.
(b) The camera was facing east.

E1

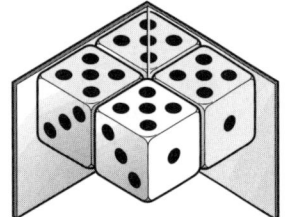

You would see 28 spots altogether.

E2 (a)

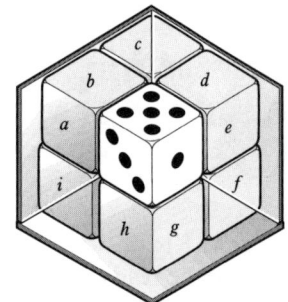

Face	Number of spots
a	3
b	5
c	5
d	5
e	1
f	1
g	1
h	3
i	3

(b)

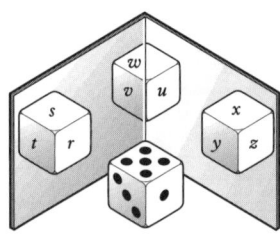

Face	Number of spots
r	6
s	5
t	3
u	6
v	4
w	5
x	5
y	4
z	1

(c)

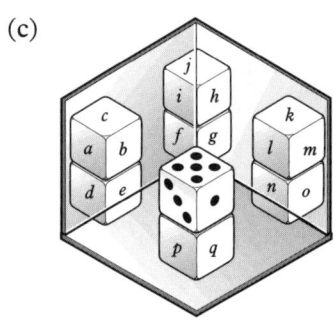

Face	Number of spots
a	3
b	6
c	5
d	3
e	6
f	4
g	6
h	6
i	4
j	5
k	5
l	4
m	1
n	4
o	1
p	3
q	1

Try this . . .
Comb your hair in a mirror, then try to comb it whilst looking at your head on a television screen through a video camera. You may be surprised! (You could, perhaps, try this in drama.)

Enlargement 1

A1

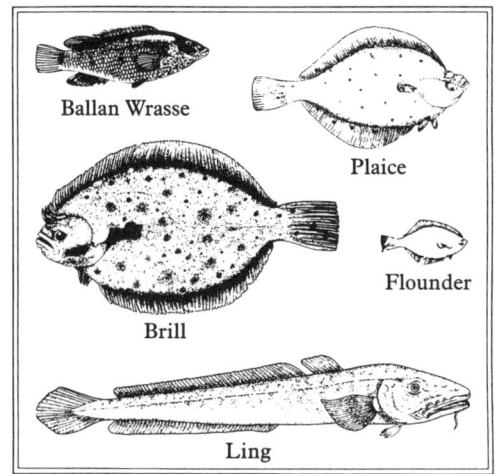

Ballan Wrasse

Plaice

Flounder

Brill

Ling

The ling is about 100 cm long so:
the ballan wrasse is about 40 cm long,
the plaice is about 50 cm long,
the brill is about 70 cm long,
and the flounder is about 20 cm long.

A2 ▲ A fake, B real, C fake, D real,
E fake, F real

How did you decide which was which?

A3 Frame B is the right shape.

Could you explain to someone why B is the right shape?

A4

(a)

(b)   (c)

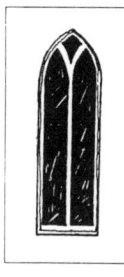

(d)   (e)   (f)

(g)

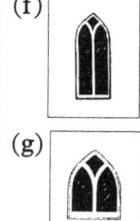

(a) Window 2   (b) Window 1
(c) Window 3   (d) Window 1
(e) Window 4   (f) Window 2
(g) Window 3

A5 (a) Window 3   (b) Window 2
(c) Window 4   (d) Window 5
(d) Window 6   (e) Window 6
(g) Window 1

B1–B5 Show your worksheets to your teacher.

B6

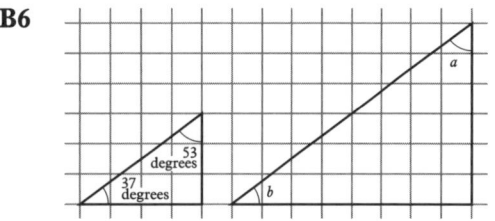

(a) Angle *a* is 53 degrees.
(b) Angle *b* is 37 degrees.

B7 (a) 4·5 cm from your own copy of the drawing.
(b) Your own enlargment.
(c) The diagonal of the enlarged rectangle should be 9·0 cm long.

B8

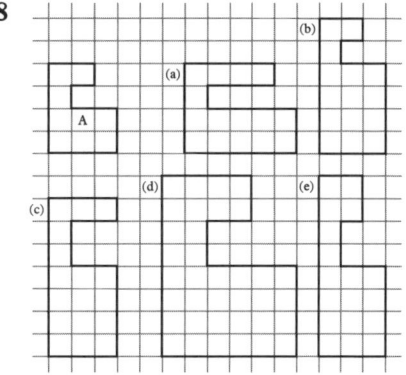

(d) is the **only** shape that is an enlargement of shape A.

29

Enlargement 2

A1 and **A2** Show your answers to your teacher.

A3

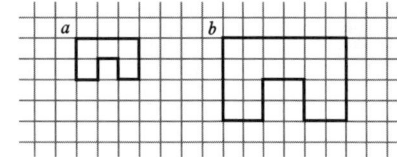

Picture *b* is a 2 times enlargement of *a*.
The scale factor of the enlargement is 2.

A4

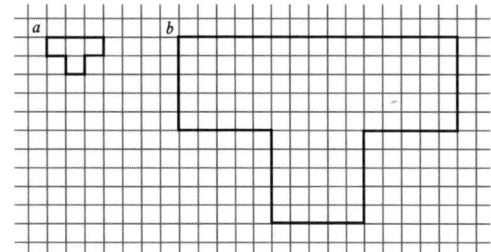

b is an enlargement of *a*. The scale factor is 5.

A5

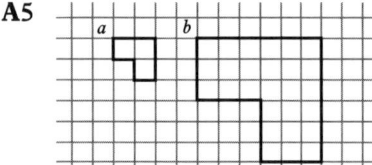

b is an enlargement of *a*. The scale factor is 3.

B1 (a) The wingspan is 4 cm.
    (b) The wingspan is 12 cm.
    (c) The scale factor is 3.

B2 Here is a completed table for the two butterflies on page 5.

OA = 2 cm	OA' = 6 cm
OB = 3 cm	OB' = 9 cm
OC = 1 cm	OC' = 3 cm
OD = 2 cm	OD' = 6 cm

B3 Here is the completed table for the two butterflies on page 6.
(The larger one is a 3 times enlargement of the smaller one.)

OA = 1 cm	OA' = 3 cm
OB = 3 cm	OB' = 9 cm
OC = 1.5 cm	OC' = 4.5 cm
OD = 2 cm	OD' = 6 cm

What do you notice about these numbers?

B4 Here is the completed table for the two leaves on page 7.

(a)

OA = 4 cm	OA' = 8 cm
OB = 4 cm	OB' = 8 cm
OC = 5 cm	OC' = 10 cm
OD = 3 cm	OD' = 6 cm
OE = 4 cm	OE' = 8 cm
OF = 3 cm	OF' = 6 cm
OG = 2 cm	OG' = 4 cm

(b) **The scale factor is 2.** How could you convince someone that this was the correct answer?

Worksheet 3-18

Check your tables from the answers given here, and then show your worksheet to your teacher.

B5

OA = 1 cm	OA' = 3 cm
OB = 2 cm	OB' = 6 cm
OC = 2 cm	OC' = 6 cm
OD = 3 cm	OD' = 9 cm

B6

OE = 2 cm	OE' = 6 cm
OF = 1 cm	OF' = 3 cm
OG = 3 cm	OG' = 9 cm

B7

OP = 2.5 cm	OP' = 5.0 cm
OQ = 3.3 cm	OQ' = 6.6 cm
OR = 2.7 cm	OR' = 5.4 cm

B8 Show your enlargement to your teacher.
▲

B9

OA = 2 cm	OA' = 6 cm
OB = 3 cm	OB' = 9 cm
OC = 4 cm	OC' = 12 cm

B10

OD = 4.4 cm	OD' = 8.8 cm
OE = 6.7 cm	OE' = 13.4 cm
OF = 2.5 cm	OF' = 5.0 cm

B11

OP = 2.5 cm	OP' = 10 cm
OQ = 3 cm	OQ' = 12 cm
OR = 2.2 cm	OR' = 8.8 cm

C1

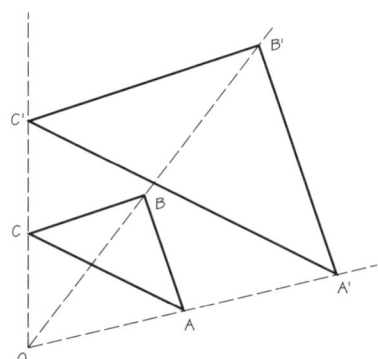

A (4, 1)	A' (8, 2)
B (3, 4)	B' (6, 8)
C (0, 3)	C' (0, 6)

C2

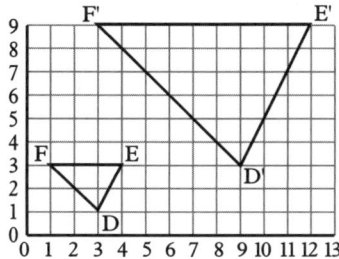

D (3, 1)	D' (9, 3)
E (4, 3)	E' (12, 9)
F (1, 3)	F' (3, 9)

D1 Show your picture to your teacher.

D2

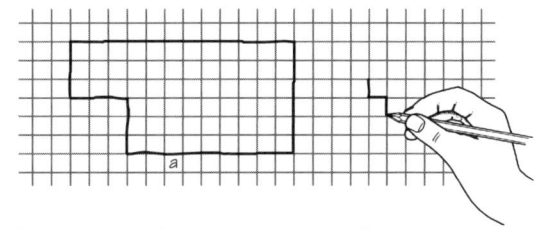

Show your picture to your teacher.
You should have written: *Picture b is a $\frac{1}{3}$ times reduction of picture a.*

D3 Show your picture to your teacher.
▲ You should have written: *picture b is a $\frac{1}{4}$ times reduction of picture a.*

Worksheet 3-20
Check your tables, then show your worksheet to your teacher.

D4

OA = 7 cm	OA' = 3.5 cm
OB = 8 cm	OB' = 4 cm
OC = 6 cm	OC' = 3 cm

D5

OD = 12 cm	OD' = 4 cm
OE = 6 cm	OE' = 2 cm
OF = 10.5 cm	OF' = 3.5 cm

D6

OP = 8 cm	OP' = 2 cm
OQ = 6 cm	OQ' = 1.5 cm
OR = 12 cm	OR' = 3 cm
OS = 4 cm	OS' = 1 cm

D7 (a) 13 cm long (b) 1·2 cm long

Enlargement 2: extension

Worksheet 3-21

Check your tables from the answers given here, then show your worksheet to your teacher.

A1

$OA = 2.7\,cm$	$OA' = 5.4\,cm$
$OB = 2.9\,cm$	$OB' = 5.8\,cm$
$OC = 2.2\,cm$	$OC' = 4.4\,cm$

A2

$OP = 3\,cm$	$OP' = {}^-6\,cm$
$OQ = 5\,cm$	$OQ' = {}^-10\,cm$
$OR = 4\,cm$	$OR' = {}^-8\,cm$

A3, A4 and **A5**

▲ Show your diagrams to your teacher.

A6 This diagram shows a negative enlargement.

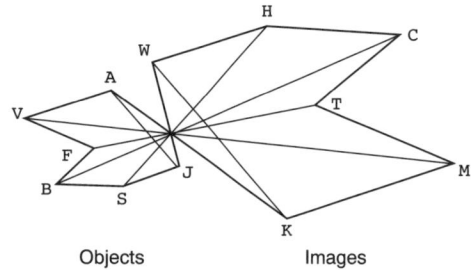

Objects   Images

  (a) K is the image of A.
  (b) C is the image of B.
  (c) W is the image of J.
  (d) H is the image of S.
  (e) T is the image of F.
  (f) M is the imge of V.

A7

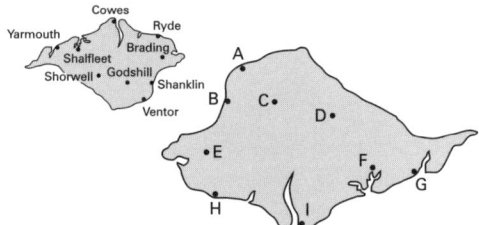

A	Ventnor	B	Shanklin
C	Godshill	D	Shorwell
E	Brading	F	Shalfleet
G	Yarmouth	H	Ryde
I	Cowes		

A8 (a) OA is 2 cm.   (b) OA' is 10 cm.
    (c) The scale factor for the enlargement is $^-5$.

A9

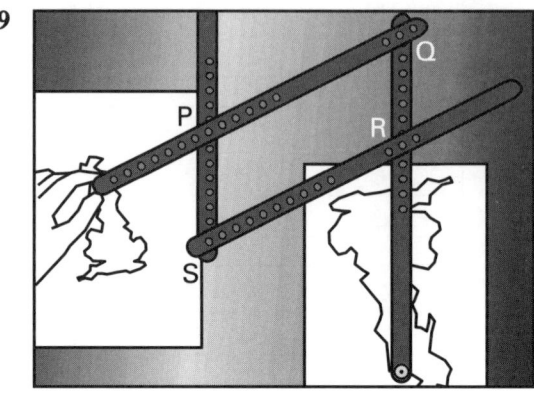

  (a) S is the centre of enlargement.
  (b) The scale factor is approximately $^-2$.

A10 This is supposed to show a $^-2$ enlargement.

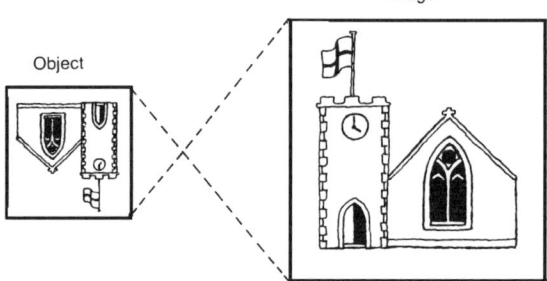

The flag, the hour hand of the clock and the door opening are all on the wrong side.

A11

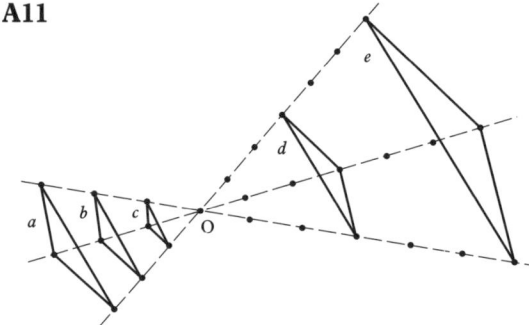

  (a) e is a $\times 2$ enlargement of triangle d.
  (b) If triangle c is the object and d its image, there has been a $^-3$ times scale factor of enlargement.

A12

Object	Image	Scale factor
a	e	⁻2
b	e	⁻3
c	e	⁻6
c	a	3
a	d	⁻1

A13

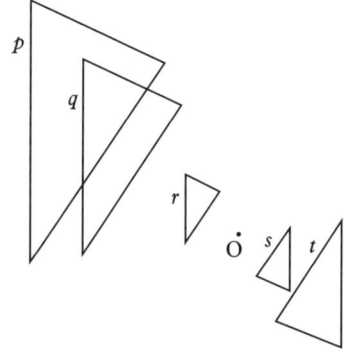

Triangle *t* is a ×2 enlargement of triangle *s*.

A14 (a) *q* is a ×3 enlargement of *r*.
  (b) *q* is a ×⁻3 enlargement of *s*.

A15 *t* is a ×⁻2 enlargement of *r*.
  p is a ×⁻2 enlargement of *t*.

B1

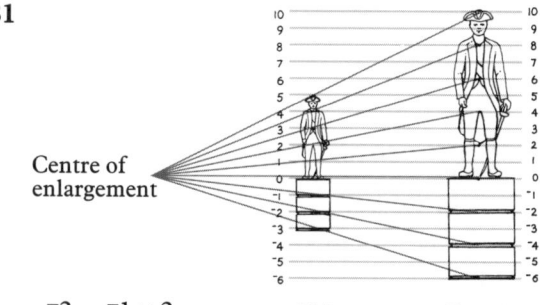

$$-2 = -1 \times 2$$
  Objects   Images

B2 (a) $-3 \times 2 = -6$ (b) $-4 \times 2 = -8$
  (c) $-6 \times 2 = -12$ (d) $-10 \times 2 = 20$

B3 (a) $-5 \times 2 = -10$ (b) $-4 \times 3 = -12$
  (c) $-7 \times 4 = -28$ (d) $-5 \times 6 = -30$

B4 (a) $-8 \times 2 = -16$ (b) $-5 \times 3 = -15$
  (c) $-3 \times 7 = -21$ (d) $-10 \times 12 = -120$
  (e) $-11 \times 6 = -66$ (f) $-27 \times 3 = -81$

B5 (a) $-1\frac{1}{2} \times 2 = -3$
  (b) $-1\cdot4 \times 2 = -2\cdot8$
  (c) $-\frac{1}{2} \times 2 = -1$
  Could you explain to someone how to work
  these out?

B6 (a) ⁻8 (b) ⁻6 (c) ⁻4 (d) ⁻2 (e) 0

B7 6

B8 8

B9 (a) 10 (b) 16 (c) 20 (d) 40

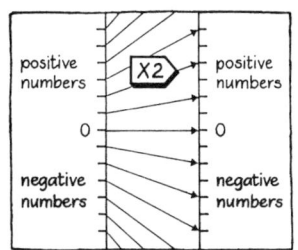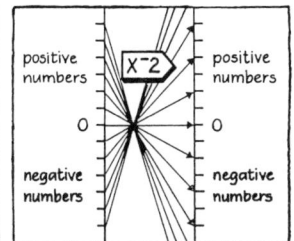

B10 (a) $-4 \times 2 = -8$
  (b) $-4 \times -2 = 8$
  (c) $4 \times -2 = -8$
  (d) $4 \times 2 = 8$
  (e) $6 \times -2 = -12$
  (f) $-6 \times -2 = 12$
  (g) $-6 \times 2 = -12$
  (h) $-3 \times -2 = 6$
  (i) $-2 \times -2 = 4$
  (j) $-7 \times 2 = -14$
  (k) $-5 \times -2 = 10$
  (l) $5 \times -2 = -10$

B11 (a) 15 (b) ⁻15 (c) ⁻15 (d) 15
  (e) ⁻24 (f) 24 (g) ⁻24 (h) ⁻40
  (i) ⁻40 (j) 40 (k) ⁻40 (l) 42

C1

(a) P (2, 2), Q(2, ⁻1), R(⁻1, ⁻1), S(⁻1, 2)
(b) Under a × 2 enlargement whose centre
  is (0, 0) the image is P'Q'R'S' where P'
  is (4, 4), Q' is (4, ⁻2), R' is (⁻2, ⁻2) and
  S' is (⁻2, 4).
(c) and (d) Your own checks.

33

C2 (a)

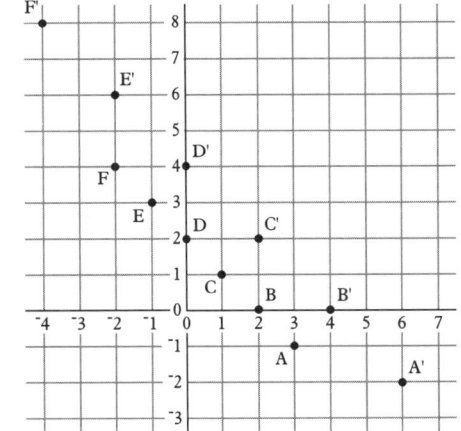

(b) A' (6, ⁻2), B' (4, 0), C' (2, 2),
D' (0, 4), E' (⁻2, 6), F' (⁻4, 8)

C3 (a) (1, ⁻2), (1, 2), (2, 2), (0, 3),
(⁻2, 2), (⁻1, 2), (⁻1, ⁻2)

(b) (⁻2, 4), (⁻2, ⁻4), (⁻4, ⁻4), (0, ⁻6),
(4, ⁻4), (2, ⁻4), (2, 4)

(c)

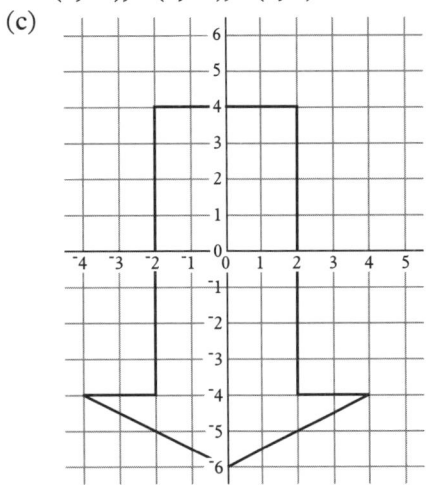

C4 Show your enlargement to your teacher.

C5 (a), (b) and (c)

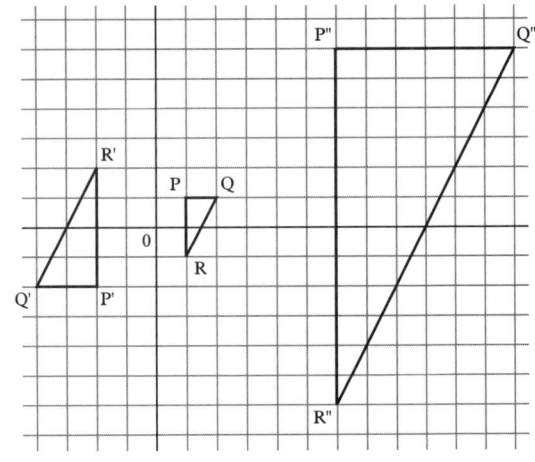

(d) The scale factor is 6.

Rotation symmetry

A1 These letters look the same upside down.

X O S H

A2 (a) Only the word 'OXO' looks the same in a mirror.

(b) All three words 'OXO', 'pod' and 'NOON' look the same upside down.

A3 Your own words which look the same upside down but not the same in a mirror.

A4 OX looks the same in a mirror placed above it, but does not look the same upside down.
Did you find any other words?

A5 ▲ You should be able to check your own answers, but just in case . . .

(a)

Looks the same upside down.

(b)

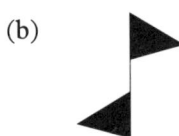

Looks the same upside down.

(c)

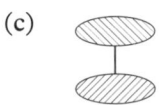

Does not look the same upside down, (look carefully at the way the lines slope).

(d)

Looks the same upside down.

(e)

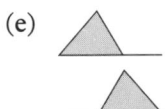

Does not look the same upside down.

(f)

Does not look the same upside down.

(g)

Looks the same upside down.

(h)

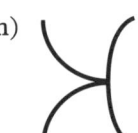

Does not look the same upside down.

A6 Here are some more dates which look the same upside down.

1691   1001   9696   69   6889

You probably found some others.

A7 You should already have checked each other's work.

Investigate

What did you both find out about your set of playing cards. Can you think why playing cards appear to look the same whichever way up you hold them?

B1 ▲ (a)

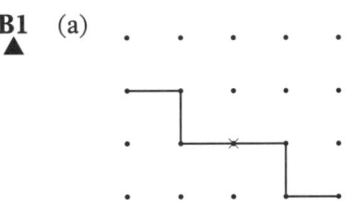

Rotation symmetry order 2.

(b)

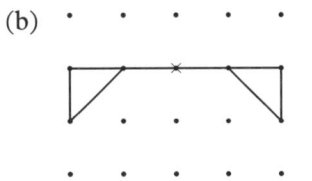

No rotation symmetry.

(c)

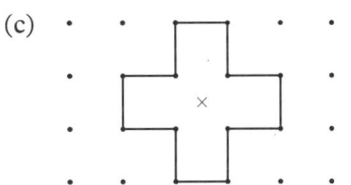

Rotation symmetry order 4.

B2 Here is how **B1**(b) and (c) may be altered to have rotation symmetry of order 2.
There are lots of other ways.
You may need to check your own answers.

(a)

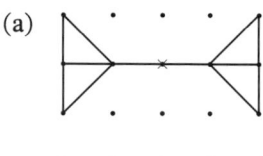

(b)

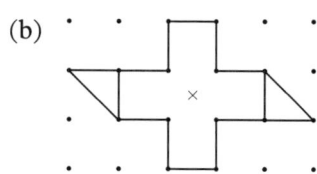

B3 All these shapes have rotation symmetry of order 2.

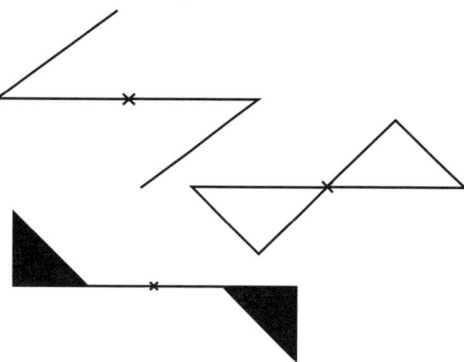

B4 This shape has rotation symmetry of order 3.

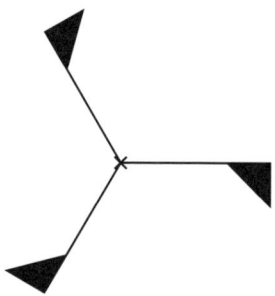

C1 The centres of rotation of these shapes are
▲ marked with an 'X'.

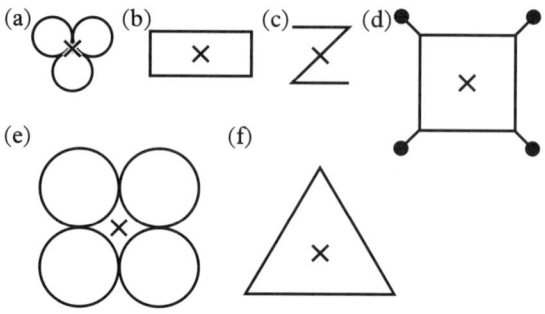

C2 (a)

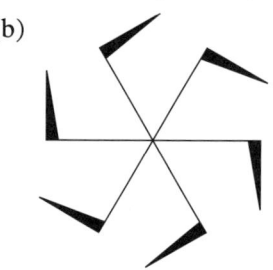

Rotation symmetry of order 4

(b)

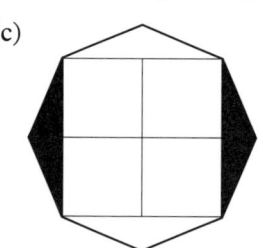

Rotation symmetry of order 6

(c)

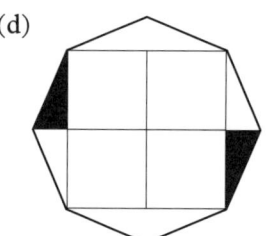

Rotation symmetry of order 2

(d)

Rotation symmetry of order 2

C3 You may have shaded your shapes slightly
differently.
See your teacher if you're not quite sure.

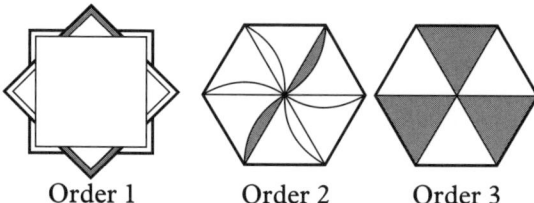

Order 1   Order 2   Order 3

D1 (a) The order of rotation symmetry is
equal to the number of equal sections
in each of the shapes.
(b) There are 360° in a complete turn.
So one-sixth of a turn will be
$360° \div 6 = 60°$.

D2 Your own patterns with rotation symmetry
of order 3.

D3 Your own patterns which have rotation
symmetry. Make sure they are labelled.
Show your patterns to your teacher.

Investigate

What about this shape?

Z

What order rotation symmetry does it
have?
How many lines of reflection symmetry
does it have?
(If you have time look at some of the
shapes in *Mirrors and reflection 1* page 10.)

Solids, mirrors and turning

A1

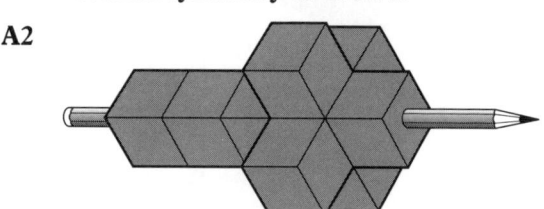

You should have found that there were two
positions where your model looked the
same. In other words the model has
rotation symmetry of order 2.

A2

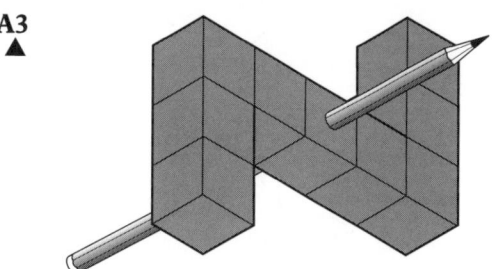

The solid has rotation symmetry order 4.
If you cannot see why, then ask your
teacher.

A3
▲

(a) This solid has rotation symmetry of
order 2.

(b) Here are some ways two other cubes may be added so that the order of rotation symmetry is still 2.

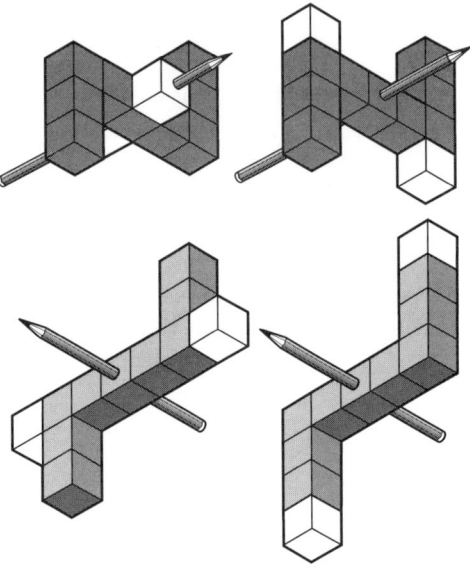

(c) Here is one way that three extra cubes can be added so that the new solid still has rotation symmetry order 2.
You should be able to find some others.

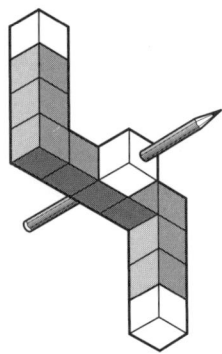

A4 This solid has rotation symmetry order 2.

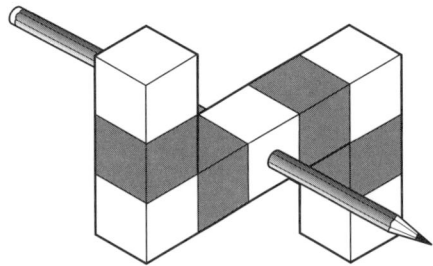

A5

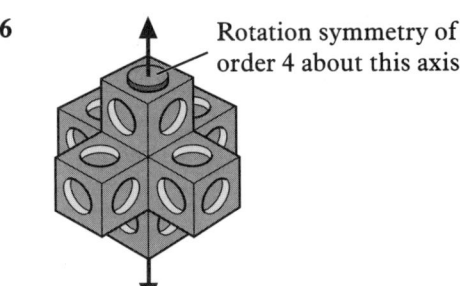

The solid has rotation symmetry of order 4 about both axis 1 and axis 2.

A6

Rotation symmetry of order 4 about this axis.

If you do not ignore the holes and connectors, the rotation symmetry of the solid changes. In **A5** the solid had the same order of rotation symmetry about all three axes of rotation (1, 2 and 3). This is not true when you build the real shape from multilink. You can build the shape so that there is one axis about which the solid has rotation symmetry of order 4. You can also build it so that there are no axes of rotation symmetry. Can you see how?

A7 Here are some other axes of rotation symmetry.

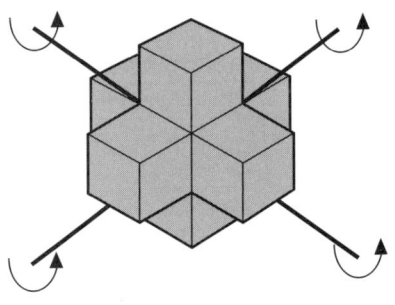

A8

(a)  (b) (c)

(d) (e)

(f) (g)

(a) A square-based pyramid has an axis of rotation symmetry as shown. The order of rotation symmetry is 4.

(b) Some scissors have an axis of rotation symmetry as shown where the order of rotation symmetry is 2. What about any scissors you looked at?

(c) A cone has an axis of rotation symmetry as shown. What is the order of rotation symmetry?

(d) Most binoculars do not have an axis of rotation symmetry.

(e) This prism has an axis of rotation symmetry as shown. The order of rotation symmetry is 3. (You may have found some axes of order 2 as well.)

(f) The cup has no axis of rotation symmetry. How about the saucer?

(g) The rotation symmetry of the vase will depend upon the pattern on it.

A9

When a key has rotation symmetry order 2 along its length it means that the motorist can open or start the car whichever way up the key is.

B1 (a) He is quite wrong, as you can see.

(b) Does this solid have a plane of symmetry? Does it have rotation symmetry?

(c) What about the thread! Does it have rotation symmetry?

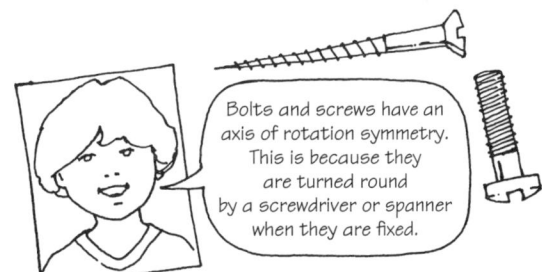

(d) This is true. For example with a hexagonal bolt head there are six possible positions which a spanner can fit. (In a circle there would be a very very large number, but you could not turn the bolt!)

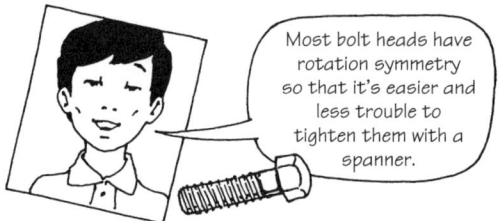

Most bolt heads have rotation symmetry so that it's easier and less trouble to tighten them with a spanner.

(e) If you ignore any writing on the pencil and assume that it has been sharpened into a perfect cone it has rotation symmetry order 6. (Your answer may depend on the shape of pencil you were thinking about.)

An ordinary pencil has just one axis of rotation symmetry. It has rotation symmetry of order three.

Try some of these

- There are three axes of rotation symmetry through the centres of sides. These all have order four.

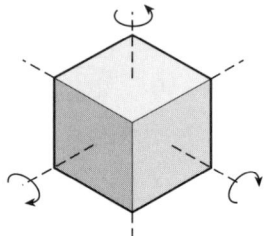

There are four axes of rotation symmetry through vertices. These all have order three.

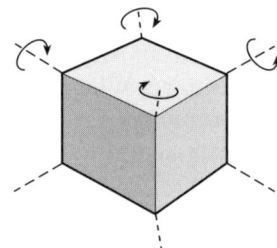

- (a) It's easy to make a solid from nine multilink cubes which has no axes of rotation symmetry.
  (b) This solid has three axes of rotation symmetry.

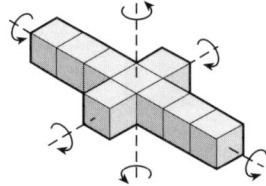

You may have found some others.

(c) This solid has one axis of rotation symmetry order 2.
You may have found some others.

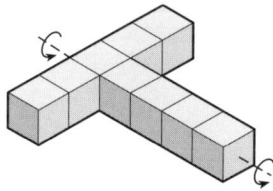

- A regular octahedron has nine planes of symmetry.
Three of them are of this type.

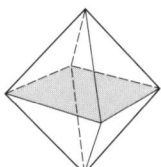

Six of them are of this type.

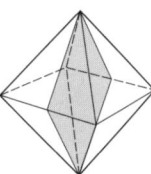

Symmetry

A1 (a)

Three lines of reflection symmetry

(b)

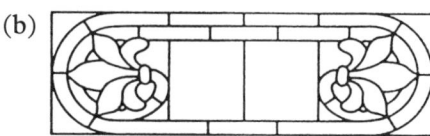

One line of reflection symmetrty

(c)

No lines of reflection symmetry

(d)

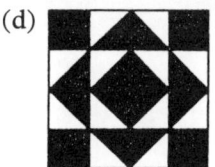

Four lines of reflection symmetry

(e)

Two lines of reflection symmetry

A2 (a), (d) and (e) also have rotation symmetry.

The terms **2-fold rotation**, 3-fold rotation ... are the same as **rotation symmetry of order 2**, rotation symmetry of order 3 ...

A3 (a)

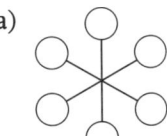

Rotation symmetry order 6

(b)

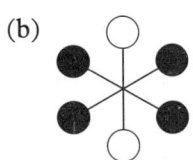

Rotation symmetry order 2

(c)

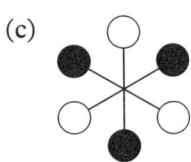

Rotation symmetry order 3

Remember

The symbol for a 2-fold centre is ⬤ (or ⬤)
The symbol for a 3-fold centre is ▲ .
The symbol for a 4-fold centre is ■ .

A4

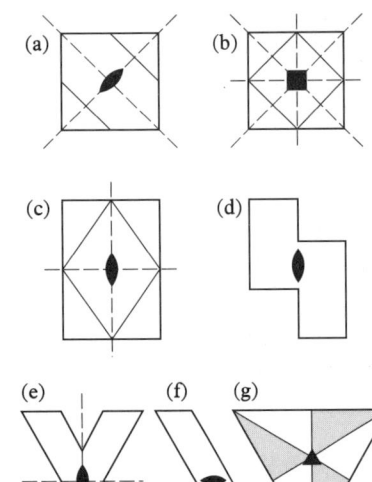

B1 Check that your patterns really are different.
These two are really the same because the second is a rotation of the first.

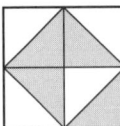 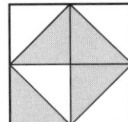

Now show your patterns to your teacher.

B2 You cannot make a pattern with two lines of reflection symmetry and no rotation symmetry.

B3

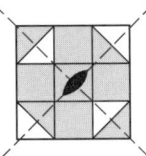

B4

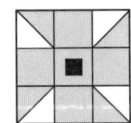

B5

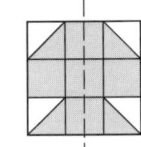

B6

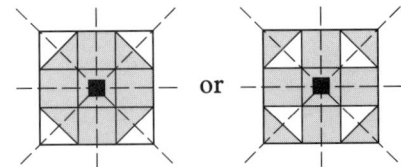

B7 You can only make two different patterns (see the answer to **B6**).
Both of them have rotation symmetry order 4 (or a 4-fold rotation centre).

B8 Show your patterns to your teacher.

B9 Show your patterns to your teacher.

B10
(a) (b) (c) (d)

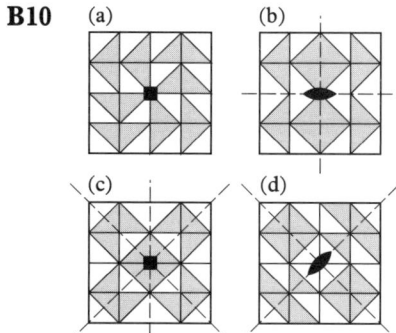

B11

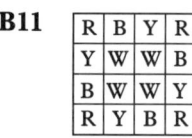

R	B	Y	R
Y	W	W	B
B	W	W	Y
R	Y	B	R

B12 Show your pattern to your teacher.

B13 Show your patterns to your teacher.

C1 You have to move your tracing 4 cm along before it fits the pattern again.

C2 We say that the period of the pattern is 4cm.

C3 Here are the patterns, beside each one is its period.

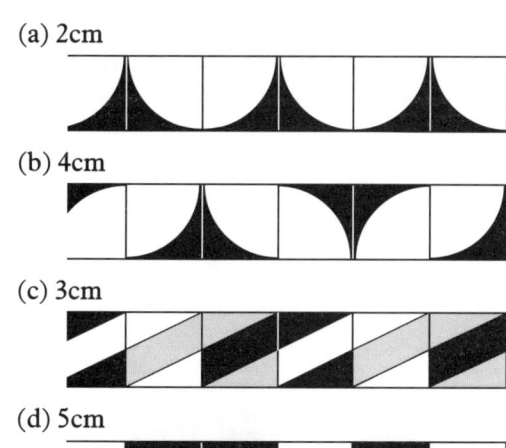

(a) 2cm

(b) 4cm

(c) 3cm

(d) 5cm

(e) 8cm

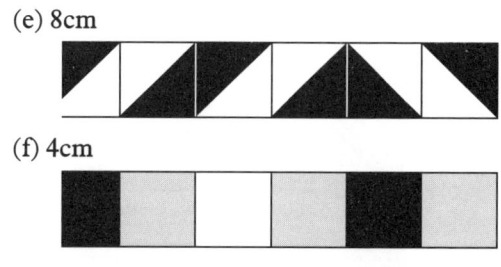

(f) 4cm

C4

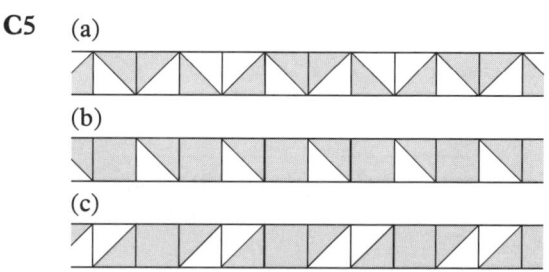

The pattern has a period (or repeat distance) of 3·4 cm.

C5 (a)

(b)

(c)

C6 ▲ There are 5 vertical lines of symmetry between these two dotted lines.

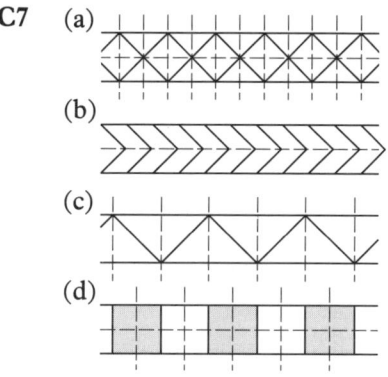

C7 (a)

(b)

(c)

(d)

43

C8 (a)

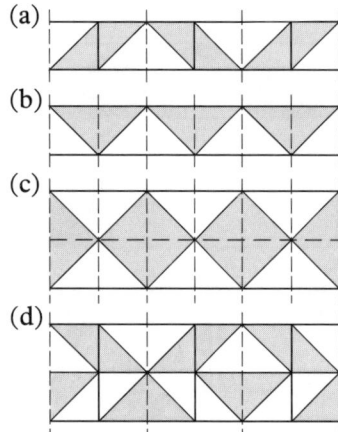

(b)

(c)

(d)

C9 See the inside back cover.

C10 (a)

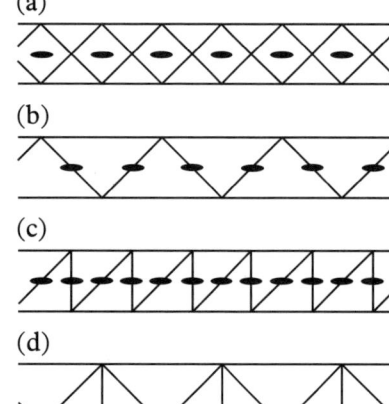

(b)

(c)

(d)

C11 Show worksheet 4-13 to your teacher.

D1

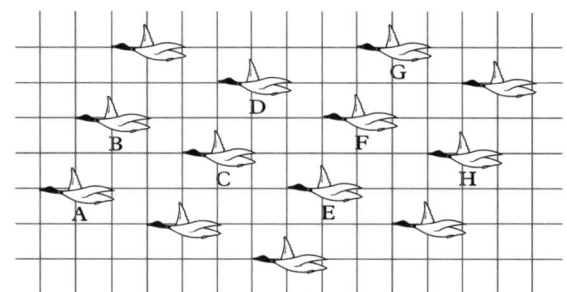

If you translated the whole design so that
A fitted onto C then:
(a) duck B would fit over duck D,
(b) duck E would fit over duck H.

D2

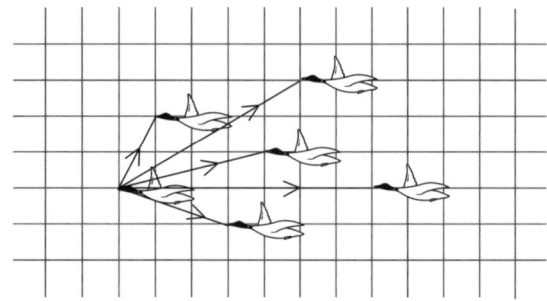

The column vectors for each of these shifts
(in order, starting from the top and going
clockwise) are $\begin{bmatrix} 1 \\ 2 \end{bmatrix}$, $\begin{bmatrix} 5 \\ 3 \end{bmatrix}$, $\begin{bmatrix} 4 \\ 1 \end{bmatrix}$, $\begin{bmatrix} 7 \\ 0 \end{bmatrix}$, $\begin{bmatrix} 3 \\ -1 \end{bmatrix}$.

D3 Any of these column vectors will translate
the pattern onto itself.

$\begin{bmatrix} 2 \\ -3 \end{bmatrix}$, $\begin{bmatrix} -1 \\ -2 \end{bmatrix}$, $\begin{bmatrix} -4 \\ -1 \end{bmatrix}$, $\begin{bmatrix} -3 \\ 1 \end{bmatrix}$, $\begin{bmatrix} -2 \\ 3 \end{bmatrix}$

There are many others.

D4

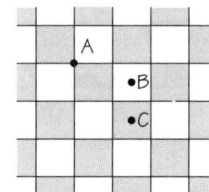

(a) A is a 2-fold rotation centre (or
rotation centre of order 2).
(b) B is a 4-fold rotation centre (or
rotation centre of order 4).
(c) C is a 4-fold rotation centre (or centre
of rotation of order 4).

D5 Show your diagram to your teacher.

D6

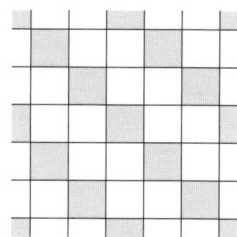

    (a) There is a 2-fold rotation centre where two green squares touch.

    (b) There is a 2-fold rotation centre inside a green (shaded) square.

    (c) There is not a centre of rotation of order 2 or 4 inside a white square.

    (d) There is a 2-fold centre where two white squares meet.

Show your drawing to your teacher.

D7 Show worksheet 4-14 to your teacher.

D8 Show worksheet 4-15 to your teacher.

Location

Maps, plans and grids — Digging into history — Bearings and journeys — Vectors 1 — Vectors 2

Starting fractions 1

Movement

Dots, lines and networks — Solids, mirrors and turning — Symmetry

Mirrors and reflection 1 — Reflection 2 — Rotation symmetry — Reflection puzzles

Enlargement 1 — Enlargement 2 — Enlargement 2: extension

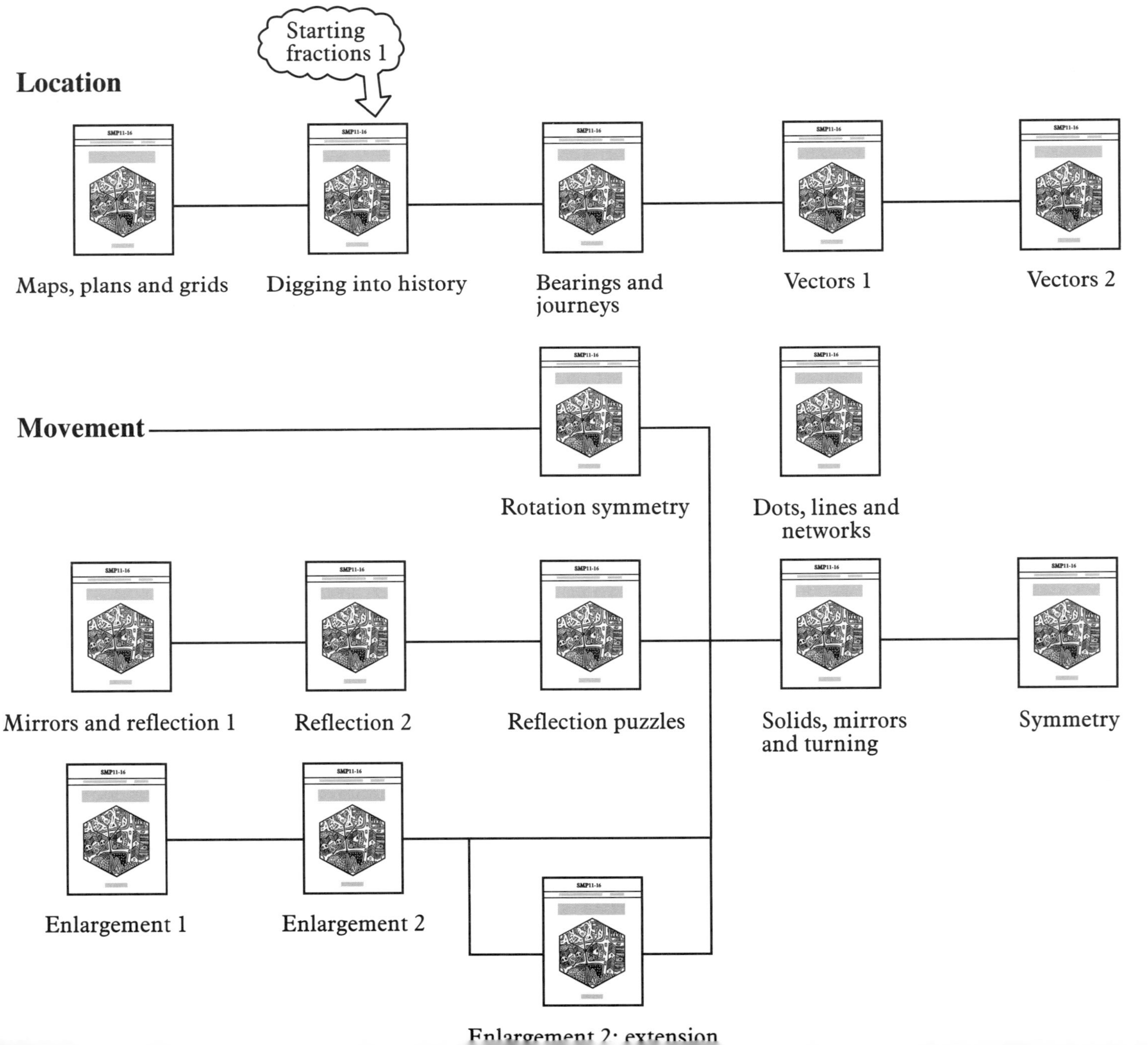